Math Mammoth
Grade 7-B Worktext

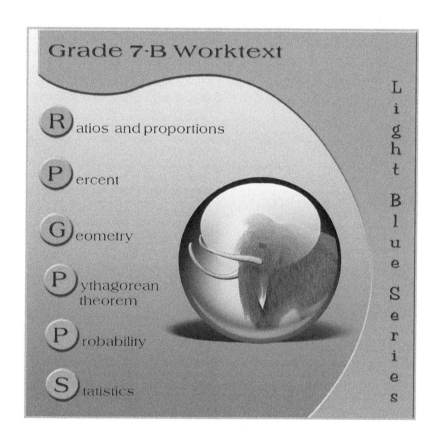

By Maria Miller

Photo credits:

p. 163 Louvre; photo by Alvesgaspar; licensed under CC BY_SA3.0

p. 217, map of Nashville © OpenStreetMap contributors
Licensed under the license at www.openstreetmap.org/copyright

Copyright 2015 - 2019 Maria Miller
ISBN 978-1511769914

Edition 1/2019

All rights reserved. No part of this book may be reproduced or transmitted in any form or by any means, electronic or mechanical, or by any information storage and retrieval system, without permission in writing from the author.

Copying permission: For having purchased this book, the copyright owner grants to the teacher-purchaser a limited permission to reproduce this material for use with his or her students. In other words, the teacher-purchaser MAY make copies of the pages, or an electronic copy of the PDF file, and provide them at no cost to the students he or she is actually teaching, but not to students of other teachers. This permission also extends to the spouse of the purchaser, for the purpose of providing copies for the children in the same family. Sharing the file with anyone else, whether via the Internet or other media, is strictly prohibited.

No permission is granted for resale of the material.

The copyright holder also grants permission to the purchaser to make electronic copies of the material for back-up purposes.

If you have other needs, such as licensing for a school or tutoring center, please contact the author at
https://www.MathMammoth.com/contact.php

Contents

Foreword	6

Chapter 6: Ratios and Proportions

Introduction	7
Ratios and Rates	12
Solving Problems Using Equivalent Rates	15
Solving Proportions: Cross Multiplying	17
Why Cross-Multiplying Works	23
Unit Rates	24
Proportional Relationships	29
Graphing Proportional Relationships–More Practice	35
More on Proportions	37
Scaling Figures	41
Floor Plans	47
Maps	51
Significant Digits	57
Chapter 6 Mixed Review	59
Chapter 6 Review	62

Chapter 7: Percent

Introduction	67
Review: Percent	72
Solving Basic Percentage Problems	75
Percent Equations	78
Circle Graphs	83
Percentage of Change	85
Percentage of Change: Applications	88
Comparing Values Using Percentages	92
Simple Interest	96
Chapter 7 Mixed Review	102
Chapter 7 Review	105

Chapter 8: Geometry

Introduction	107
Angle Relationships	113
Angles in a Triangle	118
Basic Geometric Constructions	123
More Constructions	129
Drawing Problems	134
Circumference of a Circle	141
Area of a Circle	144
Proving the Formula for the Area of a Circle	147
Area and Perimeter Problems	149
Surface Area	154
Conversions Between Customary Units of Area	158
Conversions Between Metric Units of Area	161
Slicing Three-Dimensional Shapes	164
Volume of Prisms and Cylinders	171
Chapter 8 Mixed Review	175
Chapter 8 Review	178

Chapter 9: Pythagorean Theorem

Introduction	186
Square Roots	189
Equations That Involve Taking a Square Root	193
The Pythagorean Theorem	198
The Pythagorean Theorem: Applications	203
A Proof of the Pythagorean Theorem	210
Chapter 9 Mixed Review	211
Chapter 9 Review	214

Chapter 10: Probability

Introduction	217
Probability	222
Probability Problems from Statistics	225
Experimental Probability	227
Counting the Possibilities	230
Using Simulations to Find Probabilities	236
Probabilities of Compound Events	242
Chapter 10 Mixed Review	246
Chapter 10 Review	249

Chapter 11: Statistics

Introduction 251
Review of Data Analysis 255
Random Sampling 260
Using Random Sampling 264
Comparing Two Populations 271
Comparing Two Samples 279
Chapter 11 Mixed Review 285
Chapter 11 Review 289

Foreword

Math Mammoth Grade 7 comprises a complete math curriculum for the seventh grade mathematics studies. This is a pre-algebra course, so students can continue to an algebra 1 curriculum after studying this.

The curriculum meets and actually exceeds the Common Core Standards (CCS) for grade 7. The two major areas where it exceeds those standards are linear equations (chapter 5) and the Pythagorean Theorem (chapter 9). Linear equations are covered in more depth than the CCS requires, and the Pythagorean Theorem belongs to grade 8 in the CCS. You can access a document detailing the alignment information either on the Math Mammoth website or in the download version of this curriculum.

The main areas of study in Math Mammoth Grade 7 are:

- The basics of algebra (expressions, equations, inequalities, graphing);
- Integers;
- Ratios, proportions, and percent;
- Geometry;
- Probability and statistics.

This book, 7-B, covers rations and proportions (chapter 6), percent (chapter 7), geometry (chapter 8), the Pythagorean Theorem (chapter 9), probability (chapter 10), and statistics (chapter 11). The rest of the topics are covered in the 7-A worktext.

Some important points to keep in mind when using the curriculum:

- The two books (parts A and B) are like a "framework", but you still have a lot of liberty in planning your student's studies. The five chapters in part 7-A are best studied in the order presented. However, you can study the chapters on geometry, probability, and statistics at most any point during the year. The chapters on ratios & proportions and percent (in part 7-B) are best left until the student has learned to solve one-step equations (in chapter 3).

  Math Mammoth is mastery-based, which means it concentrates on a few major topics at a time, in order to study them in depth. However, you can still use it in a *spiral* manner, if you prefer. Simply have your student study in 2-3 chapters simultaneously. This type of flexible use of the curriculum enables you to truly individualize the instruction for the student.

- Don't automatically assign all the exercises. Use your judgment, trying to assign just enough for your student's needs. You can use the skipped exercises later for review. For most students, I recommend to start out by assigning about half of the available exercises. Adjust as necessary.

- For review, the curriculum includes a worksheet maker (Internet access required), mixed review lessons, additional cumulative review lessons, and the word problems continually require usage of past concepts. Please see more information about review (and other topics) in the FAQ at
  https://www.mathmammoth.com/faq-lightblue.php

I heartily recommend that you view the full user guide for your grade level, available at
https://www.mathmammoth.com/userguides/

Lastly, you can find free videos matched to the curriculum at https://www.mathmammoth.com/videos/

> *I wish you success in teaching math!*
>
> *Maria Miller, the author*

Chapter 6: Ratios and Proportions
Introduction

Chapter 6 reviews the concept, which has already been presented in previous grades, of the ratio of two quantities. From this concept, we develop the related concepts of a rate (so much of one thing per so much of another thing) and a proportion (an equation of ratios). We also study how tables of equivalent ratios can help to solve problems with rates, and how cross-multiplying can help to solve problems with proportions.

The lesson *Unit Rates* defines the concept of the unit rate, shows how to calculate one, and gives practice at doing so, including practice with complex fractions. We also consider rates as two quantities that vary, graph the corresponding equation in the coordinate grid, and tie in the concept of unit rate with the concept of slope.

The concept of direct variation is introduced in the lesson *Proportional Relationships*. Writing and graphing equations gives a visual understanding of proportionality. In two following lessons, students also practice solving rate problems in different ways, using the various methods they have learned throughout the chapter.

The lessons *Scaling Figures, Floor Plans, and Maps* give useful applications and more practice to master the concepts of proportions.

Before the *Chapter Review* there is also an optional lesson, *Significant Digits*, that deals with the concept of the accuracy of a measurement and how it limits the accuracy of the solution. It is optional because significant digits is not a standard topic for seventh grade, yet the concept in it is quite important, especially in science.

Keep in mind that the specific lessons in the chapter can take several days to finish. They are not "daily lessons." Instead, use the general guideline that seventh graders should finish about 12 pages a week in order to finish the curriculum in about 40 weeks. Also, I recommend not assigning all the exercises by default, but that you use your judgment, and strive to vary the number of assigned exercises according to the student's needs.

Please see the user guide at https://www.mathmammoth.com/userguides/ for more guidance on using and pacing the curriculum.

There are free videos matched to the curriculum at https://www.mathmammoth.com/videos/ (choose 7th grade).

The Lessons in Chapter 6

	page	span
Ratios and Rates	12	*3 pages*
Solving Problems Using Equivalent Rates	15	*2 pages*
Solving Proportions: Cross Multiplying	17	*6 pages*
Why Cross-Multiplying Works	23	*1 page*
Unit Rates	24	*5 pages*
Proportional Relationships	29	*6 pages*
Graphing Proportional Relationships–More Practice	35	*2 pages*
More on Proportions	37	*4 pages*
Scaling Figures	41	*6 pages*
Floor Plans	47	*4 pages*
Maps	51	*6 pages*
Significant Digits	57	*2 pages*
Chapter 6 Mixed Review	59	*3 pages*
Chapter 6 Review	62	*5 pages*

Helpful Resources on the Internet

Ratios and Proportions—Video Lessons by Maria
A set of free videos about ratios, rates, and proportions—by the author.
http://www.mathmammoth.com/videos/proportions.php

RATIOS AND RATES

Ratio Pairs Matching Game
Match cards representing equivalent ratios.
Easy: http://nrich.maths.org/4824   Challenge: http://nrich.maths.org/4821

All About Ratios - Quiz
An interactive five-question quiz about equivalent ratios presented with pictures
http://math.rice.edu/~lanius/proportions/quiz1.html

Rate Module from BrainingCamp
A comprehensive interactive lesson on the concepts of ratio, rate, and constant speed (for 6th and 7th grades). Includes an animated lesson, a virtual manipulative, and questions and problems to solve.
http://www.brainingcamp.com/lessons/rates/

Ratios Activity from BBC Bitesize
An animated tutorial about dividing in a given ratio and scale models with some quiz questions along the way.
http://www.bbc.co.uk/education/guides/znnycdm/activity

Ratios and Rates: Quick Quiz
Practice ratios and rates with this short interactive online quiz.
http://wps.pearsoned.com.au/mfqld1/0,7860,759792--759794,00.html

Ratio Quiz from BBC Skillswise
A multiple-choice quiz about the concept of ratio. You can take the quiz online or download it as a PDF or doc file.
http://www.bbc.co.uk/skillswise/quiz/ma19rati-e1and2-quiz

Ratio Quiz from Syvum
A 10-question online quiz about ratios and problem solving.
http://www.syvum.com/cgi/online/mult.cgi/gmat/math_review/arithmetic_5.tdf?0

Unit Rates Involving Fractions
Practice computing rates associated with ratios of fractions or decimals in this interactive activity.
https://www.khanacademy.org/math/cc-seventh-grade-math/cc-7th-ratio-proportion/cc-7th-rates/e/rate_problems_1

Ratio Word Problems
Reinforce your ratios skills with these interactive word problems.
https://www.khanacademy.org/math/pre-algebra/pre-algebra-ratios-rates/pre-algebra-ratio-word-problems/e/ratio_word_problems

Comparing Rates
Practice completing rate charts in this interactive online activity.
https://www.khanacademy.org/math/pre-algebra/pre-algebra-ratios-rates/pre-algebra-rates/e/comparing-rates

Free Ride
An interactive activity about bicycle gear ratios. Choose the front and back gears, which determines the gear ratio. Then choose a route, pedal forward, and make sure you land exactly on the five flags.
http://illuminations.nctm.org/ActivityDetail.aspx?ID=178

Exploring Rate, Ratio and Proportion (Video Interactive)
The video portion of this resource illustrates how these math concepts play a role in photography. The interactive component allows students to explore ratio equivalencies by enlarging and reducing images.
http://www.learnalberta.ca/content/mejhm/index.html?l=0&ID1=AB.MATH.JR.NUMB&ID2=AB.MATH.JR.NUMB.RATE

Three-Term Ratios
Practice the equivalency of ratios by filling in the missing numbers in three-term ratios (for example, 2:7:5 = __ : 105 : ___) where the numbers represent the amounts of three colors in different photographs. Afterwards you get to assemble a puzzle from the nine photographs.
http://www.learnalberta.ca/content/mejhm/index.html?l=0&ID1=AB.MATH.JR.NUMB&ID2=AB.MATH.JR.NUMB.RATE&lesson=html/object_interactives/3_term_ratio/use_it.html

If the two links above don't work, use this link:
http://www.learnalberta.ca/content/mejhm/index.html?l=0&ID1=AB.MATH.JR.NUMB
First choose Rate/Ratio/Proportion, and then either *Exploring Rate, Ratio, and Proportion* or *3-Term Ratios*.

PROPORTIONS

Ratios and Proportions
A tutorial with interactive practice exercises about ratios and proportions.
https://www.wisc-online.com/learn/formal-science/mathematics/gem2004/ratios-and-proportions

Solving Proportions Practice
In this interactive practice, you can choose to show a hint "vertically", "horizontally", or algebraically.
http://www.xpmath.com/forums/arcade.php?do=play&gameid=97

Solving Proportions
Practice solving basic proportions with this interactive exercise from Khan Academy.
https://www.khanacademy.org/exercise/proportions_1

Ratios, Rates, and Proportions Quiz
Use this multiple-choice self-check quiz to test your knowledge about ratios, rates, and proportions.
http://www.phschool.com/webcodes10/index.cfm?wcprefix=ara&wcsuffix=0504&area=view

Proportions: Short Quiz
Use this multiple-choice self-check quiz to test your knowledge about ratios, rates, and proportions.
http://www.phschool.com/webcodes10/index.cfm?wcprefix=bja&wcsuffix=0602&area=view

Challenge Board
Choose questions from the challenge board about rates, ratios, and proportions.
http://www.quia.com/cb/158527.html
http://www.quia.com/cb/101022.html

Write a Proportion to a Problem
Practice writing proportions to describe real-world situations in this interactive exercise.
https://www.khanacademy.org/math/pre-algebra/pre-algebra-ratios-rates/pre-algebra-write-and-solve-proportions/e/writing_proportions

Proportion Word Problems
Practice setting up and solving proportions to solve word problems in this interactive online activity.
https://www.khanacademy.org/math/pre-algebra/pre-algebra-ratios-rates/pre-algebra-write-and-solve-proportions/e/constructing-proportions-to-solve-application-problems

Rags to Riches—Proportions
Solve proportions and advance towards more and more difficult questions.
http://www.quia.com/rr/35025.html

Pennies to Heaven
This activity uses pennies to give students a context to investigate large numbers and measurements.
https://www.illustrativemathematics.org/content-standards/6/RP/A/3/tasks/1291

PROPORTIONAL RELATIONSHIPS

Gym Membership Plans
You are shown two situations and asked which one represents a proportional relationship. Click "Student view" to print the student worksheet. The solution is included further down the web page.
https://www.illustrativemathematics.org/content-standards/7/RP/A/2/tasks/1983

Proportional Relationships
A short video showing how ratios, tables, and graphs can help identify proportional relationships. A worksheet is also available.
https://www.pbslearningmedia.org/resource/muen-math-rp-proportionalrelationships/proportional-relationships/

Proportional Relationships
Practice telling whether or not the relationship between two quantities is proportional by reasoning about equivalent ratios.
https://www.khanacademy.org/math/pre-algebra/pre-algebra-ratios-rates/pre-algebra-proportional-rel/e/analyzing-and-identifying-proportional-relationships

Proportional Relationships: Graphs
Practice telling whether or not the relationship between two quantities is proportional by looking at a graph of the relationship.
https://www.khanacademy.org/math/cc-seventh-grade-math/cc-7th-ratio-proportion/cc-7th-graphs-proportions/e/analyzing-and-identifying-proportional-relationships-2

Interpreting Graphs of Proportional Relationships
Practice reading and analyzing graphs of proportional relationships in this interactive online exercise.
https://www.khanacademy.org/math/cc-seventh-grade-math/cc-7th-ratio-proportion/cc-7th-equations-of-proportional-relationships/e/interpreting-graphs-of-proportional-relationships

Writing Proportional Equations
Practice writing equations to describe proportional relationships in this interactive online activity.
https://www.khanacademy.org/math/cc-seventh-grade-math/cc-7th-ratio-proportion/cc-7th-equations-of-proportional-relationships/e/writing-proportional-equations

SCALE DRAWINGS AND MAPS

Similar Triangles
A short lesson dealing with similar triangles and calculating the ratio of corresponding sides. At the bottom of the page, there is a set of practice questions.
https://www.mathsisfun.com/geometry/triangles-similar.html

Similar Shapes Exercises
Answer questions about the scale factors of lengths, areas, and volumes of similar shapes.
http://www.transum.org/Maths/Activity/Similar/

Scale Drawings Exercise
Measure line segments and angles in geometric figures, including interpreting scale drawings.
http://www.transum.org/software/Online_Exercise/ScaleDrawing/

Scale Drawings Quizzes
Interactive self-check quizzes about scale drawings. By reloading the page you will get different questions.
http://www.phschool.com/webcodes10/index.cfm?wcprefix=ara&wcsuffix=0506&area=view

https://www.thatquiz.org/tq/practicetest?sy5p7cgwt2nf

Ratio and Scale
An online unit about scale models, scale factors, and maps with interactive exercises and animations.
http://www.absorblearning.com/mathematics/demo/units/KCA024.html

Use Proportions to Solve Problems Involving Scale Drawings
A set of word problems. You can choose how they are presented: as flashcards, as a quiz where you match questions and answers, as a multiple choice quiz, or a true/false quiz. You can also play a game (Jewels).
http://www.cram.com/flashcards/use-proportions-to-solve-problems-involving-scale-drawings-3453121

Scale Drawings - Problem Solving and Constructing Scale Drawings Using Various Scales
A comprehensive lesson with several worked out examples concerning scale drawings.
http://www.ck12.org/user:c2ZveDJAb3N3ZWdvLm9yZw../book/Oswego-City-School-District---Grade-7-Common-Core/section/12.0/

Constructing Scale Drawings
Practice making scale drawings on an interactive grid. The system includes hints and the ability to check answers.
https://www.khanacademy.org/math/cc-seventh-grade-math/cc-7th-geometry/cc-7th-scale-drawings/e/constructing-scale-drawings

Interpreting Scale Drawings
Solve word problems involving scale drawings in an online practice environment.
https://www.khanacademy.org/math/cc-seventh-grade-math/cc-7th-geometry/cc-7th-scale-drawings/e/interpreting-scale-drawings

Scale Drawings and Maps Quiz
Answer questions about scales on maps and scale drawings in these five self-check word problems.
http://www.math6.org/ratios/8.6_quiz.htm

Maps
A tutorial with worked out examples and interactive exercises about how to calculate distances on the map or in real life based on the map's scale.
http://www.cimt.org.uk/projects/mepres/book7/bk7i19/bk7_19i3.htm

Short Quiz on Maps
Practice map-related concepts in this multiple-choice quiz.
http://www.proprofs.com/quiz-school/story.php?title=map-scales

SIGNIFICANT DIGITS

Sig Fig Rules
Drag Sig J. Fig to cover each significant digit in the given number.
http://www.sigfig.dreamhosters.com/

Practice on Significant Figures
A multiple-choice quiz that also reminds you of the rules for significant digits.
https://web.archive.org/web/20161128075136/http://www.chemistrywithmsdana.org/wp-content/uploads/2012/07/SigFig.html

Significant digits quiz
A 10-question multiple-choice quiz about significant digits.
http://www.quia.com/quiz/114241.html?AP_rand=1260486279

Ratios and Rates

A **ratio** is a comparison of two numbers, or quantities, using division.

For example, to compare the hearts to the stars in the picture, we say that the ratio of hearts to stars is 5:10 (read "five to ten").

The two numbers in the ratio are called the **first term** and the **second term** of the ratio. The order in which these terms are mentioned does matter! For example, the ratio of stars to hearts is *not* the same as the ratio of hearts to stars. The former is 10:5 and the latter is 5:10.

We can write this ratio in several different ways:

- The ratio of hearts to stars is 5:10.
- The ratio of hearts to stars is 5 to 10.
- The ratio of hearts to stars is $\frac{5}{10}$.
- For every five hearts, there are ten stars.

Note that we are not comparing two numbers to determine which one is greater (as in 5 < 10). The comparison is relative as in a multiplication problem. For example, the ratio 5:10 can be simplified to 1:2, and it indicates to us that there are twice as many stars as there are hearts.

We **simplify ratios** in exactly the same way we simplify fractions.

Example 1. In the picture at the right, the ratio of hearts to stars is 12:16. We can simplify that ratio to 6:8 and even further to 3:4. These three ratios (12:16, 6:8, and 3:4) are called **equivalent ratios**.

The ratio that is simplified to lowest terms, 3:4, tells us that for every three hearts, there are four stars.

1. Write the ratio and then simplify it to lowest terms.

    The ratio of triangles to diamonds is _____ : _____ = _____ : _____ .

    In this picture, there are _____ triangles to every _____ diamonds.

2. **a.** Draw a picture with pentagons and circles so that the ratio of pentagons to the total of all the shapes is 7:9.

   b. What is the ratio of circles to pentagons?

3. **a.** Draw a picture in which (1) there are three diamonds for every five triangles, and (2) there is a total of 9 diamonds.

   b. Write the ratio of all the diamonds to all the triangles, and simplify this ratio to lowest terms.

4. Write the equivalent ratios.

a. 5 to 45 = 1 to _____	**b.** 3 : _____ = 9 : 60	**c.** 280 : 420 = 2 : _____	**d.** $\frac{5}{13} = \frac{}{65}$

> We can also form **ratios using quantities that have units**. If the units are the same, they cancel.
>
> **Example 2.** Simplify the ratio 250 g : 1.5 kg.
>
> First we convert 1 kg to grams and then simplify: $\dfrac{250 \text{ g}}{1.5 \text{ kg}} = \dfrac{250 \text{ g}}{1{,}500 \text{ g}} = \dfrac{250}{1{,}500} = \dfrac{1}{6}$.

5. Use a fraction line to write ratios of the given quantities as in the example. Then simplify the ratios.

a. 5 kg and 800 g $\dfrac{5 \text{ kg}}{800 \text{ g}} =$	b. 600 cm and 2.4 m
c. 1 gallon and 3 quarts	d. 3 ft 4 in and 1 ft 4 in

> We can generally **convert ratios with decimals or fractions into ratios of whole numbers**.
>
> **Example 3.** Because we can multiply both terms of the ratio by 10, $\dfrac{1.5 \text{ km}}{2 \text{ km}} = \dfrac{15 \text{ km}}{20 \text{ km}}$.
>
> Then: $\dfrac{15 \text{ km}}{20 \text{ km}} = \dfrac{15}{20} = \dfrac{3}{4}$. So the ratio 1.5 km : 2 km is equal to 3:4.
>
> You can also see that the ratio is 3:4 by noticing that both 1.5 km and 2 km are evenly divisible by 500 m.
>
> **Example 4.** Simplify the ratio ¼ mile to 5 miles.
>
> First, ¼ mi : 5 mi = ¼ : 5. Multiplying both terms of the ratio by 4, we get ¼ : 5 = 1:20.

6. Use a fraction line to write ratios of the given quantities. Then simplify the ratios to integers.

a. 5.6 km and 3.2 km	b. 0.02 m and 0.5 m
c. 1.25 m and 0.5 m	d. 1/2 L and 7 1/2 L
e. 1/2 mi and 3 1/2 mi	f. 2/3 km and 1 km

If the two terms in a ratio have *different* units, then the ratio is also called a **rate**.
Example 5. The ratio "5 miles to 40 minutes" is a rate that compares the quantities "5 miles" and "40 minutes," perhaps for the purpose of giving us the speed at which a person is running. We can write this rate as 5 miles : 40 minutes or $\frac{5 \text{ miles}}{40 \text{ minutes}}$ or 5 miles *per* 40 minutes. The word "per" in a rate signifies the same thing as a colon or a fraction line.
This rate can be simplified: $\frac{5 \text{ miles}}{40 \text{ minutes}} = \frac{1 \text{ mile}}{8 \text{ minutes}}$. The person runs 1 mile in 8 minutes.
Example 6. Simplify the rate "15 pencils per 100¢." Solution: $\frac{15 \text{ pencils}}{100¢} = \frac{3 \text{ pencils}}{20¢}$.

7. Write each rate using a colon, the word "per," or a fraction line. Then simplify it.

   a. Jeff swims at a constant speed of 1,200 ft in 15 minutes.

   b. A car can travel 54 miles on 3 gallons of gasoline.

8. Fill in the missing numbers to form equivalent rates.

a. $\frac{1/2 \text{ cm}}{30 \text{ min}} = \frac{}{1 \text{ h}} = \frac{}{15 \text{ min}}$	**b.** $\frac{\$88.40}{8 \text{ hr}} = \frac{}{2 \text{ hr}} = \frac{}{10 \text{ hr}}$

9. Simplify these rates. Don't forget to write the units.

a. 280 km per 7 hours	**b.** 2.5 inches : 1.5 minutes

10. A car is traveling at a constant speed of 72 km/hour. Fill in the table of equivalent rates: each pair of numbers in the table (distance/time) forms a rate that is equivalent to the rate 72 km/hour.

Distance (km)							
Time (min)	10	30	40	50	60	90	100

11. Eight pairs of socks cost $20. Fill in the table of equivalent rates.

Cost ($)								
Pairs of socks	1	2	4	6	7	8	9	10

Solving Problems Using Equivalent Rates

Example 1. It took Jack 1 1/2 hours to paint 24 feet of fence. Painting at the same speed, how long will it take him to paint the rest of the fence, which is 100 feet long?

In this problem, we see a rate of 24 ft per 1 1/2 hours. There is another rate, too: 100 ft per an unknown amount of time. These two are equivalent rates. We can use a table of equivalent rates to solve the problem.

Amount of fence (ft)	24	4	48	96	100
Time (minutes)	90	15	180	360	375

(1) We figure that Jack can paint 4 ft of fence in 15 minutes (by dividing the terms in the original rate by 6).

(2) Next we double both terms in the original rate of 24 ft/90 min to get the rate 48 ft/180 min.

(3) Then we double that rate to get 96 ft/360 min.

(4) Lastly, since 100 ft = 96 ft + 4 ft, we add the corresponding times to get
360 min + 15 min = 375 min, or 6 hours 15 minutes.

Example 2. You get 20 erasers for $1.90. How much would 22 erasers cost?

We can solve the problem using a table of equivalent ratios:

Price	$1.90	$0.19	$2.09
Erasers	20	2	22

Twenty erasers cost $1.90, so 2 erasers would cost 1/10 that much, or $0.19. Lastly, we find the cost of 22 erasers, which is the cost of 20 + 2 erasers, or $1.90 + $0.19 = $2.09.

1. Fill in the tables of equivalent rates.

a. Distance	15 km			
Time	3 hr	1 hr	15 min	45 min

b. Pay	$6				
Time	45 min	15 min	1 hr	1 hr 45 min

2. Fill in the missing terms in these equivalent rates.

a. $\dfrac{3 \text{ in}}{8 \text{ ft}} = \dfrac{}{2 \text{ ft}} = \dfrac{}{12 \text{ ft}} = \dfrac{}{20 \text{ ft}}$

b. $\dfrac{115 \text{ words}}{2 \text{ min}} = \dfrac{}{1 \text{ min}} = \dfrac{}{3 \text{ min}}$

3. Jake can ride his bicycle 8 miles in 28 minutes. At the same constant speed, how long will he take to go 36 miles? Fill in the equivalent rates below.

$\dfrac{8 \text{ miles}}{14 \text{ minutes}} = \dfrac{4 \text{ miles}}{ \text{ minutes}} = \dfrac{36 \text{ miles}}{ \text{ minutes}}$

4. Larry earns $90 for seven hours of work. In how many hours will he earn $600?

Earnings							
Work Hours							

> **Example 3.** Jake can ride his bike 20 miles in 45 minutes. Riding at the same constant speed, how far could he ride in 1 hour?
>
> Remember that we can <u>multiply or divide both terms of a rate</u> by the same number to form another, equivalent rate. You have used this same idea in the past with equivalent fractions.
>
>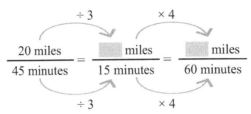
>
> It is not easy to go directly from the rate of 20 miles in 45 minutes to the equivalent rate for 60 minutes. However, we can easily find the rate for 15 minutes: simply divide both 20 miles and 45 minutes by 3. In case you are stumped by 20 ÷ 3, remember that it is easy to solve when you think of it as a fraction: 20/3 = 6 2/3. We get the rate 6 2/3 miles per 15 minutes.
>
> Then, we multiply both terms of that rate by 4. Again, don't be intimidated by the fraction: 4 · (6 2/3) = 4 · 6 + 4 · (2/3) = 24 + 8/3 = 26 2/3. So Jake can ride 26 2/3 miles in 1 hour.

5. A car can travel 45 miles on 2 gallons of gasoline. How many gallons of gasoline would the car need for a trip of 60 miles?

6. **a.** A train travels at a constant speed of 80 miles per hour. How far will it go in 140 minutes?

    $$\frac{80 \text{ miles}}{60 \text{ min}} =$$

   b. Is this equal to the rate of traveling 50 miles in 40 minutes?

7. You get 30 pencils for $4.50. Is that equal to the rate of 50 pencils for $7.25? You can make a table of equivalent rates to help yourself.

Price							
Pencils							

8. In a poll that interviewed 1,000 people about their favorite color, 640 people said they liked blue.

    a. Simplify this ratio to lowest terms.

    b. Assuming the same ratio holds true in another group of 125 people, how many of those people can we expect to like blue?

Solving Proportions: Cross Multiplying

A **proportion** is an equation where one ratio is set equal to another ratio.

The equations below are examples of proportions. Notice that, in each one, one ratio is equal to another.

$$\frac{20 \text{ km}}{30 \text{ min}} = \frac{55 \text{ km}}{x} \qquad \frac{7}{8} = \frac{21}{24} \qquad \frac{12.2}{T} = \frac{15.6}{8.7}$$

You can solve proportions in several different ways. For example, you can think of them as equivalent fractions or use a table of equivalent ratios like we did in the previous lesson. However, the most common way of solving proportions is **cross-multiplying**. It is the most efficient method to use when the numbers are not simple.

Example 1. Solve the proportion below.

$\frac{7}{16} = \frac{143}{x}$ — This is the proportion.

$\frac{7}{16} \times \frac{143}{x}$ — First, **cross-multiply** (multiply crisscross) as indicated by the lines in the picture: $7 \cdot x$ and $16 \cdot 143$.
(Realize that what you are really doing is multiplying both sides by 16 and also by x, but the 16 cancels on the left and the x cancels on the right.)

$7x = 16 \cdot 143$ — After cross-multiplying we get this equation.

$7x = 2{,}288$ — Here, we have multiplied $16 \cdot 143 = 2{,}288$ on the right side. Next we will divide both sides of the equation by 7.

$\frac{7x}{7} = \frac{2{,}288}{7}$ — In this last step, the left side simplifies to x. On the right side, we simply perform the division.

$x \approx 326.86$ — This is the final answer.

You can use a calculator for all the problems in this lesson.

1. Solve the proportions step-by-step. Round your answers to the nearest tenth.

a. $\frac{15}{32} = \frac{67}{M}$ — First cross-multiply.

$15M =$ ____ — This is the equation you get after cross-multiplying.

____ = ____ — In this step, calculate what is on the right side.

____ = ____ — In this step, divide both sides of the equation by ____.

$M =$ ____ — This is the final answer.

b. $\frac{7}{146} = \frac{38}{S}$ — First cross-multiply.

____ = ____ — This is the equation you get after cross-multiplying.

____ = ____ — Calculate $146 \cdot 38$.

____ = ____ — In this step, divide both sides of the equation by ____.

$S =$ ____ — This is the final answer.

17

2. Solve the proportions step-by-step. Round your answers to the nearest tenth.

a. $\dfrac{1.2}{4.5} = \dfrac{G}{7.0}$ First cross-multiply.

(You may want to flip the sides so the variable is on the left.)

Calculate what is on the right side.

___ = ___   In this step, divide both sides of the equation by ___.

G = ___   This is the final answer.

b. $\dfrac{4.3}{C} = \dfrac{10.0}{17.0}$ First cross-multiply.

(You may want to flip the sides so the variable is on the left.)

Calculate what is on the right side.

___ = ___   In this step, divide both sides of the equation by ___.

R = ___   This is the final answer.

3. Solve the following proportions by using cross-multiplication. Give your answers to the nearest hundredth.

a. $\dfrac{T}{25} = \dfrac{15}{3}$

b. $\dfrac{17}{214} = \dfrac{2}{M}$

Example 2. An International cell phone call costs $41 per hour. How much would a call of 27 minutes cost?

The problem involves two rates: $41 per 1 hour and an unknown cost per 27 minutes. To write a proportion, use a variable for the unknown and set those two rates to be equal. Note also that we need to change the 1 hour to 60 minutes so that the two amounts of time use the same unit.

$$\frac{\$41}{60 \text{ min}} \diagup\!\!\!\!\diagdown \frac{M}{27 \text{ min}}$$
This is the proportion. We cross-multiply as indicated by the lines: 60 · M and 41 · 27.

$\$41 \cdot 27 \text{ min} = 60 \text{ min} \cdot M$
This is the equation we get after cross-multiplying. We will keep the units "min" and "$" in the equation.

$60 \text{ min} \cdot M = \$41 \cdot 27 \text{ min}$
We switched the sides to put the variable on the left.

$60 \text{ min} \cdot M = \$1107 \cdot \text{min}$
Here, we have multiplied 27 · 41 = 1107 on the right side. Notice, the unit "min" is still kept in the equation.

$$\frac{\cancel{60 \text{ min}} \cdot M}{\cancel{60 \text{ min}}} = \frac{\$1107 \cdot \cancel{\text{min}}}{60 \cancel{\text{min}}}$$
Now we divide both sides of the equation by the quantity "60 min" — not only by 60, but by "60 min." On the left side, "60 min" and "60 min" cancel out. On the right side, the units "min" cancel. Notice that the unit "$" does not cancel out, which means our final answer will be in dollars.

$M = \$18.45$
The 27-minute international cell phone call would cost $18.45.

Here is another example problem that shows you how to work with the units while solving the equation.

Example 3. A jet airplane flies 5,600 km in 4 hours 40 minutes. Maintaining that same speed, how long will it take to fly 2,150 km?

Again, the problem involves two equal rates, so we can solve it by using a proportion. First we choose the variable T for the unknown time. Next we write an equals sign between the two rates, 5,600 km per 280 minutes and 2,150 km per T. Then we also convert the flying time of 4 hours and 40 minutes into 280 minutes.

$$\frac{5,600 \text{ km}}{280 \text{ min}} = \frac{2,150 \text{ km}}{T}$$
Here is the proportion. Notice that the distances are on top and the times are on the bottom — they "match."

$$\frac{5,600 \text{ km}}{280 \text{ min}} \diagup\!\!\!\!\diagdown \frac{2,150 \text{ km}}{T}$$
Now we cross-multiply: 5,600 km · T and 2,150 km · 280 min. Notice that each of these multiplications involves two *different* quantities: a <u>distance</u> by a <u>time</u>. If we get a <u>distance</u> times a <u>distance</u> or a <u>time</u> times a <u>time</u>, then the proportion is set up incorrectly.

$5,600 \text{ km} \cdot T = 2,150 \text{ km} \cdot 280 \text{ min}$
This is what we have after cross-multiplying. We keep the units "min" and "km" in the equation.

$5,600 \text{ km} \cdot T = 602,000 \text{ km} \cdot \text{min}$
Here we have multiplied 2,150 · 280 = 602,000 on the right side. The units "km" and "min" have been placed *behind* that number, and they are still multiplied (km · min)!

$$\frac{\cancel{5,600 \text{ km}} \cdot T}{\cancel{5,600 \text{ km}}} = \frac{602,000 \cancel{\text{km}} \cdot \text{min}}{5,600 \cancel{\text{km}}}$$
Here we divide both sides of the equation by 5,600 km. On the left side, "5,600 km" and "5,600 km" cancel out. On the right side, the units "km" cancel. The unit "min" does not cancel out.

$T = 107.5 \text{ min}$
This is the final answer. The airplane takes 107.5 minutes, or 1 hour 47.5 minutes, to fly 2,150 km.

4. For each problem below, write a proportion and solve it. Carry the units through your calculation. Don't forget to check that your answer is reasonable.

a. To paint 700 m² May needs 85 liters of paint. How much paint does she need to paint 240 m²?	**b.** A car can travel 56.0 miles on 3.2 gallons of gasoline. How far can it go with 7.9 gallons of gasoline?

5. Jack and Noah each wrote a different proportion for the problem below. Solve both proportions. Carry the units through your calculations. Which proportion gives the correct answer—or are both correct?

A car travels 154 miles on 5.5 gallons of gasoline. How many gallons would it need to travel 900 miles?	
Jack's proportion: $$\frac{154 \text{ mi}}{5.5 \text{ gal}} = \frac{900 \text{ mi}}{x}$$	Noah's proportion: $$\frac{5.5 \text{ gal}}{154 \text{ mi}} = \frac{x}{900 \text{ mi}}$$

In the previous problem, you saw that two proportions that were set up differently could both yield the correct answer. How can that be? The answer lies in the equations you get after cross-multiplying: they were identical.

For word problems that call for a proportion, there are several ways to set up the proportion correctly and several ways to set it up incorrectly. You can spot the incorrect ways by two facts:

(1) You end up multiplying quantities with the same unit, such as the amount of fuel by the amount of fuel, or the price by the price.

(2) Your answer is not reasonable.

Example 4. If 7 lb of chicken costs $12, how much would 100 lb cost?

Each of the proportions below will give you a correct answer because each one of them leads to the equation with $7 \text{ lb} \cdot x$ on one side and $\$12 \cdot 100 \text{ lb}$ on the other side.

$$\frac{7 \text{ lb}}{\$12} = \frac{100 \text{ lb}}{x} \qquad \frac{\$12}{7 \text{ lb}} = \frac{x}{100 \text{ lb}} \qquad \frac{\$12}{x} = \frac{7 \text{ lb}}{100 \text{ lb}} \qquad \frac{x}{\$12} = \frac{100 \text{ lb}}{7 \text{ lb}}$$

The proportions below will *not* work. In some of them, you end up multiplying pounds by pounds. In others, you end up multiplying x by 100 lb and getting the answer $x = \$0.84$, which is not reasonable. When things like that happen, it means the that two ratios got mixed up. Always check that your answer is reasonable.

$$\frac{7 \text{ lb}}{\$12} = \frac{x}{100 \text{ lb}} \qquad \frac{100 \text{ lb}}{\$12} = \frac{x}{7 \text{ lb}} \qquad \frac{\$12}{x} = \frac{100 \text{ lb}}{7 \text{ lb}} \qquad \frac{7 \text{ lb}}{x} = \frac{100 \text{ lb}}{\$12}$$

6. Choose the proportion that is set up correctly and solve it ("s" means seconds).

a. $\dfrac{3 \text{ s}}{23 \text{ m}} = \dfrac{x}{7 \text{ s}}$

b. $\dfrac{23 \text{ m}}{3 \text{ s}} = \dfrac{x}{7 \text{ s}}$

7. Jane wrote a proportion to solve the following problem. Explain what is wrong with her proportion and correct it. Then solve the corrected proportion. Also, check that your answer is reasonable.

Twenty kilograms of premium dog food cost $51. How much would 17 kg cost?	Jane's proportion: $$\frac{20 \text{ kg}}{\$51} = \frac{x}{17 \text{ kg}}$$

8. For each problem below, write a proportion and solve it. Remember to check that your answer is reasonable.

a. A 5.0 lb bag of fertilizer covers 1,000 square feet of lawn. How much fertilizer would you need for a rectangular 30 ft by 24 ft lawn?	b. To paint 700 m² you need 85 liters of paint. How much area would 6 liters of paint cover?

Why Cross-Multiplying Works

Recall that if we multiply both sides of an equation by the same number, the two sides are still equal.

In a proportion, we have two different numbers in the denominators. We can first multiply both sides of the proportion by the one denominator and then by the other, or we can cross-multiply. Cross-multiplying is in reality just a *shortcut* for doing those two separate multiplications at the same time.

Let's solve the proportion below by multiplying both sides first by the one denominator, then by the other.

$$\frac{6}{13} = \frac{93}{T}$$ First we multiply both sides by 13.

$$\frac{6}{\cancel{13}} \cdot \cancel{13} = \frac{93}{T} \cdot 13$$ Now, on the left side, the 13 in the denominator cancels the other 13.

$$6 = \frac{93}{T} \cdot 13$$ The next step is to multiply both sides by T.

$$6T = \frac{93}{\cancel{T}} \cdot 13 \cdot \cancel{T}$$ On the right side, the T in the denominator cancels the T in the numerator.

$$6T = 93 \cdot 13$$ *This* is the step to which cross-multiplying would have brought us directly. We continue solving as usual and calculate $93 \cdot 13$.

$$6T = 1209$$ To solve for T, we need to divide both sides by 6.

$$\frac{6T}{6} = \frac{1209}{6}$$ On the left side, we simplify 6 and 6 so that only T is left. On the right side, we perform the division.

$$T = 201.5$$ This is the final answer.

Cross-multiplying is not a "magic trick" but simply a *shortcut* based on mathematical principles.

1. Solve the proportion in two ways: (a) using the shortcut of cross-multiplying, (b) using the slow way as in the example above.

a. $\dfrac{8}{1.15} = \dfrac{37}{K}$

b. $\dfrac{8}{1.15} = \dfrac{37}{K}$

Unit Rates

Remember that a rate is a ratio where the two terms have different units, such as 2 kg/$0.45 and 600 km/5 h.

In a **unit rate, the second term of the rate is one** (of some unit).

For example, 55 mi/1 hr and $4.95/1 lb are unit rates. The number "1" is nearly always omitted so those rates are usually written as 55 mi/hr and $4.95/lb.

To convert a rate into an equivalent unit rate simply divide the numbers in the rate.

Example 1. Mark can ride his bike 35 km in 1 ½ hours. What is the unit rate?

To find the unit rate, we use the principles of division by fractions to divide 35 km by 1 ½ h. The units "km" and "hours" are divided, too, and become "km per hour" or "km/hour."

$$\frac{35 \text{ km}}{1 \ 1/2 \text{ h}} = 35 \div \frac{3}{2} \text{ km/h} = 35 \times \frac{2}{3} \text{ km/h}$$

$$= \frac{70}{3} \text{ km/h} = 23 \ 1/3 \text{ km/h}.$$

We could also use decimal division:
35 km ÷ 1.5 h = 23.333... km/h.

So the unit rate is 23 ⅓ km per hour.

Example 2. A snail can slide through the mud 5 cm in 20 minutes. What is the unit rate?

Here, it is actually not clear whether we should give the unit rate as cm/min or cm/hr. Let's do both.

(1) To get the unit rate in cm/min, we simply divide 5 cm ÷ 20 min. We get the fraction 5/20. We also divide the units to get "cm/min." So we get

$$5 \text{ cm} \div 20 \text{ min} = 5/20 \text{ cm/min} = 1/4 \text{ cm/min}$$

Or use decimals: 5 cm/20 min = 25/100 cm/min

= 0.25 cm/min.

(2) For centimeters per hour, we multiply both terms of the rate by 3 to get an equivalent rate of 15 cm in 60 minutes, which is 15 cm in 1 hour.

1. Find the unit rate.

   a. $125 for 5 packages

   b. $6 for 30 envelopes

   c. $1.37 for ½ hour

   d. 2 ½ inches per 4 minutes

   e. 24 m² per 3/4 gallon

2. A person is walking 1/2 mile every 1/4 hour. Choose the correct fraction for the unit rate and simplify it.

$$\frac{\frac{1}{4}}{\frac{1}{2}} \text{ miles per hour} \quad \text{or} \quad \frac{\frac{1}{2}}{\frac{1}{4}} \text{ miles per hour}$$

3. Write the unit rate as a complex fraction, and then simplify it.

a. Lisa can make three skirts out of 5 ½ yards of material. Find the unit rate for one skirt.

b. A drink made with 30 g of powdered vegetables gives you 2 ¾ servings of vegetables. Find the unit rate for 1 g of powder.

c. Marsha walked 2 ¾ miles in 5/6 of an hour.

d. It takes Linda 2 ½ hours to make 1 ½ vases by hand.

e. There are 5,400 people living in a suburban development area of 3/8 km^2.

f. Alex paid $8.70 for 5/8 lb of nuts.

g. Elijah can finish 3/8 of a game in 7/12 of an hour.

Example 3. On rough country roads, Greg averages a speed of 30 kilometers per hour with his moped.

This 30 km/h is a rate that involves two quantities—kilometers and hours, or distance and time—which we can consider as variables. See the table.

distance (km)	0	30	60	90	120	150
time (hours)	0	1	2	3	4	5

These two variables d and t are related by the equation $d = 30t$ that we can plot on the coordinate grid:

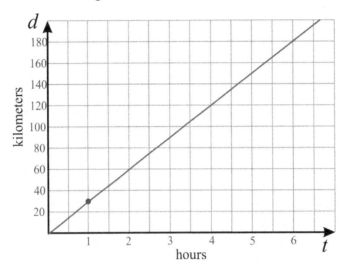

The unit rate 30 km/h is the slope of the line. It is also the coefficient of the variable t in the equation $d = 30t$.

We have plotted the point (1, 30) that matches the unit rate—1 hour and 30 kilometers.

What does the point (4, 120) mean?

It means that Greg can travel 120 kilometers in 4 hours.

4. A delivery truck is traveling at a constant speed of 50 km per hour.

   a. Write an equation relating the distance (d) and the time (t).

   b. Plot the equation you wrote in part (a) and the point that matches the unit rate.

   c. What does the point (3, 150) mean in terms of this situation?

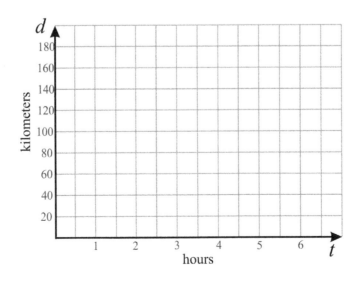

5. Some baby ducks are walking at a constant speed of 1/3 meter per second (or 1 meter in 3 seconds).

   a. Write an equation relating the distance (d) and time (t) and plot it in the grid below.

   b. What is the unit rate?

   c. Plot the point that matches the unit rate in this situation.

   d. What does the point (0, 0) mean in terms of this situation?

   e. Plot the point that matches the time $t = 4$ s.

   f. Plot the point that matches the distance $d = 3$ m.

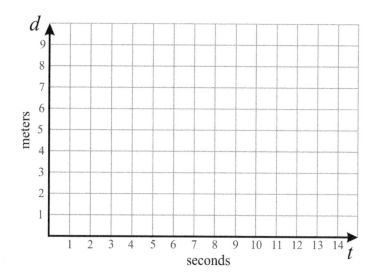

6. The equation $d = (1/2)t$ represents the distance in meters that adult ducks walk in t seconds.

   a. Plot this equation in the same grid as you did the equation for the baby ducks.

   b. Plot the point that matches the unit rate in this situation.

   c. How can you see from the graphs that the adult ducks walk faster than the babies?

   d. How much farther will the adult ducks have walked than the baby ducks at $t = 5$ s?

   e. How much longer will the baby ducks take to walk 5 meters than the adult ducks?

> **Example 4.** A town has 45,000 inhabitants and 128 doctors.
> (1) Find the number of doctors per 10,000 population.
> (2) Find the unit rate (the number of people per one doctor).
>
> (1) We can easily write the rate of doctors to all people — it is 128 doctors : 45,000 people. Since there are 4.5 groups of 10,000 people in 45,000, if we put 4.5 in place of the 45,000 in that rate and convert it to a unit rate, then we get the rate for one group of 10,000 people.
>
> So first we write the ratio 128 to 4.5, which is the ratio of doctors to 10,000 people. Then we divide 128 ÷ 4.5 = 28.444... to get the actual number of doctors for 10,000 people. So there are about 28 doctors for each 10,000 people.
>
> (2) We divide the number of people by the number of doctors: 45,000 people ÷ 128 doctors = 351.5625 people/doctor ≈ 352 people/doctor. In other words, on average, each doctor serves about 352 people.

7. **a.** Calculate the rate of physicians per 10,000 people in Bulgaria, if the country is estimated to have 27,700 doctors and 7,365,000 people. Round your answer to one decimal.

   b. Algeria has 12.1 physicians per 10,000 people. How many doctors would you expect to find in an area in Algeria that has 350,000 residents?

   c. What is the rate of physicians per 1,000 people in Algeria?

8. Jane and Stacy ran for 30 seconds. Afterward each girl checked her heartbeat. Jane counted that her heart beat 38 times in 15 seconds, and Stacy counted that her heart beat 52 times in 20 seconds.

   a. Which girl had a faster heart rate?
   How much faster?

   b. Let's say Jane keeps running and her heart keeps beating at the same rate. Write an equation for the relationship between the number of her heartbeats and time in seconds.
   Also, identify the unit rate in this situation.

   c. Do the same for Stacy.

Proportional Relationships

In this lesson we study what it means when two variables are **in direct variation** or **in proportion**. The basic idea is that whenever one variable changes, the other varies (changes) **proportionally** or **at the same rate**.

Example 1. Apples cost $3.50 per kilogram.

As you already know, "$3.50 per kilogram" is a unit rate. It relates two quantities: the cost and the weight of the apples. We will now consider those quantities to be *variables* and let them vary.

In this situation the two variables—the cost and the weight—are in **direct variation** or **in proportion**. This means that if either variable doubles, to maintain the proportion, the other must also double. If one variable increases ten times, the other also has to increase ten times. If one of them is cut to a third, the other must also be cut to a third, and so on. If you multiply or divide either variable by any number, the other variable must get multiplied or divided by the same number.

At $3.50 per kilogram, 6 kg of apples would cost $21. If I double the weight to 12 kg, the cost also doubles, to $42. If I want only 1/3 of 6 kg, or 2 kg, the cost is only 1/3 of $21, or $7.

In summary, whenever one variable in the rate changes, the other has to change proportionally.

You can check to see if two variables are in direct variation in several different ways. Here is one way.

(1) Check to see if the values of the variables are in direct variation. If you double the value of one, does the value of the other double also? If one quantity increases by 5 times, does the other do the same?

Example 2. The table gives the cost of using a computer in an internet cafe based on the number of hours of usage. Are the cost and time in proportion?

Cost ($)	3	4	5	6	7	8	9
Time (hr)	1	2	3	4	5	6	7

Look at the two rates $4 for 2 hours and $8 for 6 hours. The time triples but the cost only doubles! Therefore, the relationship between these two quantities (cost and time) is not proportional.

1. Fill in the table of values and determine whether the two variables are in direct variation.

a. $y = 3x$

y								
x	−3	−2	−1	0	1	2	3	4

b. $y = x + 2$

y								
x	−3	−2	−1	0	1	2	3	4

c. $y = (1/2)x - 1$

y								
x	−3	−2	−1	0	1	2	3	4

d. $y = -x$

y								
x	−3	−2	−1	0	1	2	3	4

e. $C = 2.4n$

C								
n	0	1	2	3	4	5	6	7

f. $h = 1/k$

h								
k	1	2	3	4	5	6	7	8

There is a basic difference between the graphs of equations that have variables in proportion and those that do not. There is also a difference in the equations themselves. Try to spot those differences in the following exercise.

2. Fill in the table of values and determine whether the two variables are in direct variation. Then plot the equations.

a. $y = x + 1$

x	−3	−2	−1	0	1	2	3	4
y								

b. $y = 2x$

x	−3	−2	−1	0	1	2	3	4
y								

c. $y = 2x - 1$

x	−3	−2	−1	0	1	2	3	4
y								

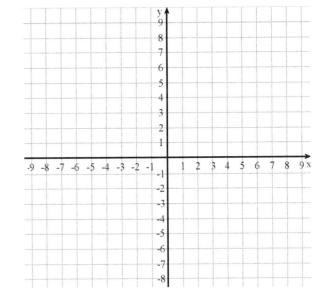

d. $y = (1/2)x$

x	−3	−2	−1	0	1	2	3	4
y								

e. $y = -2x$

x	−3	−2	−1	0	1	2	3	4
y								

f. $y = -2x + 1$

x	−3	−2	−1	0	1	2	3	4
y								

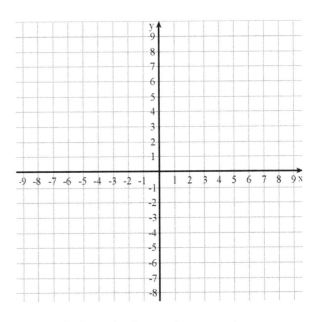

3. Now consider the plots, the equations, and the tables of values of the six items in the previous exercise. How do the equations and plots of the variables that are in proportion differ from those that aren't? If you cannot tell, check the next page.

(2) When two variables are proportional, the equation relating the two is of the form $y = mx$, where y and x are the variables, and m is a constant. The constant m is called the constant of proportionality.

For example, if apples cost $3.50 per kilogram, the equation relating the weight (w) to the cost (C) of the apples is $C = 3.5w$. The constant of proportionality is 3.5.

- The constant of proportionality is also the unit rate. When the weight of apples is 1 kg, the cost is $C = \$3.5 \cdot 1 = \$3.50$.

(3) When two quantities are proportional, their graph is a straight line that goes through the origin.

Remember, the equation relating the two quantities is of the form $y = mx$. The constant m is not only the unit rate—it is also the slope. In our apple example, the equation is $C = 3.5w$. The slope is 3.5, because whenever the weight (w-value) increases by 1 kg, the cost (C-value) increases by $3.50.

Notice the point (1, 3.5) that corresponds to the unit rate: the weight is 1 kg and the cost is $3.50.

Cost of Apples

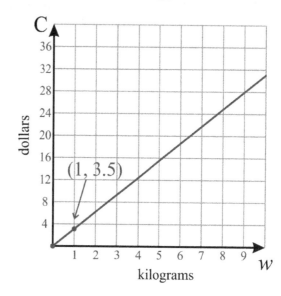

The other special point on the graph is the origin at (0, 0). That point is always on the graph because, if two quantities are in proportion, when one of them is zero, the other also has to be zero.

We graph the weight w on the horizontal axis as the independent variable and the cost C on the vertical axis as the dependent variable because we are choosing a weight of apples independently based on some need and then "observing" what its cost is. So here the cost C depends on the weight w. The dependent variable depends on the independent variable. We always plot the independent variable on the horizontal axis.

It is possible to look at the situation just the opposite way, and consider how the weight of the apples depends on the cost of the apples. In that case, we would write the equation $w = 0.2857C$ and plot the values of C on the horizontal axis. This way is not as common as observing how the cost depends on the weight.

Just one more thought: You might wonder, "Why does the line have to go through the origin if the quantities are in proportion?"

Consider the principle governing direct variation: if one quantity is cut in half, then the other is cut in half. Let's say you start with certain values of the two quantities, such as 6 meters per 2 minutes. Now cut both in half to get 3 meters per 1 minute. Do it again to get 1.5 meters per 1/2 minute. Do it again, and again.

Notice that both numbers get smaller and smaller yet—they approach zero. This would happen no matter what values of the two quantities you started with. So the point (0, 0) has to be included in the graph of quantities that are in direct variation.

4. **a.** Graph the equation $y = 3x$.

   b. State the unit rate in this situation.

   c. Plot the point that corresponds to the unit rate.

5. **a.** Graph the equation $y = 0.5x$.

   b. State the unit rate in this situation.

   c. Plot the point that corresponds to the unit rate.

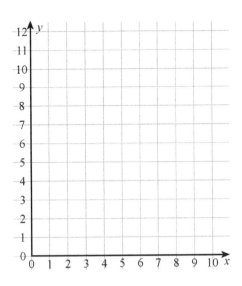

6. Determine whether the two quantities are in proportion. If so, find the unit rate, write an equation relating the two, and graph the equation.

a.

x	−3	−2	−1	0	1	2	3	4
y	−5	−4	−3	−2	−1	0	1	2

In proportion or not?

Unit rate:

Equation:

b.

x	−3	−2	−1	0	1	2	3	4
y	−12	−8	−4	0	4	8	12	16

In proportion or not?

Unit rate:

Equation:

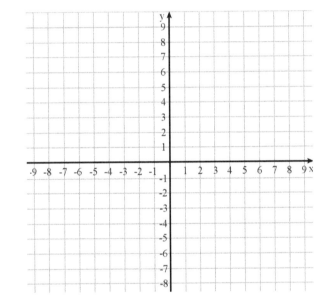

c.

x	−3	−2	−1	0	1	2	3	4
y	−1	−2/3	−1/3	0	1/3	2/3	1	4/3

In proportion or not?

Unit rate:

Equation:

d.

x	−3	−2	−1	0	1	2	3	4
y	2	5/3	4/3	1	2/3	1/3	0	−1/3

In proportion or not?

Unit rate:

Equation:

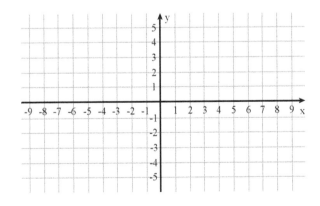

7. For each line:
   (1) state the unit rate (including the units of measurement),
   (2) plot the point that corresponds to the unit rate, and
   (3) write an equation for the line.

   a. Unit rate:

   Equation:

   b. Unit rate:

   Equation:

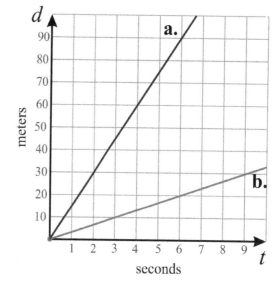

8. Write an equation for each line.

   a. Equation:

   b. Equation:

   c. What real-life situation could the equations and their plots represent?

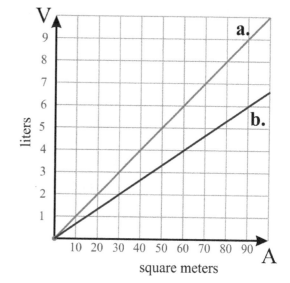

9. A car is traveling at a constant speed, covering 135 miles in three hours.

   a. Write an equation that relates the distance the car has traveled to the time it takes to do so.

   b. Plot the equation on the grid. Choose the scaling on the axes so that you can fit the point $t = 10$ hours and the corresponding distance onto the grid.

   c. Plot the point that corresponds to the unit rate.

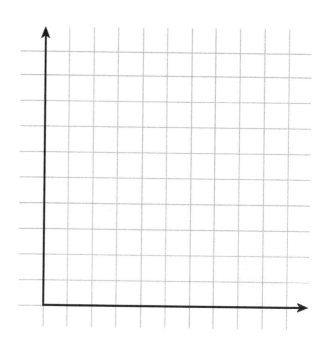

10. Robert works as a salesman. He is paid $300 per week plus a commission based on the total sales he makes. Consider Robert's pay and the number of items he sells.

Items sold in a week	0	1	2	3	4	5	6	7	8
Pay	300	350	400	450	500	550	600	650	700

   a. Are these two quantities in proportion?

   b. If so, write an equation relating the two and state the constant of proportionality.

11. Three workers planting trees on a big farm can plant 60 trees in a day, on average. Consider the number of workers and the number of trees they plant in one day.

   a. Are these two quantities in proportion?

   b. If so, write an equation relating the two and state the constant of proportionality.

12. The table shows the relationship between the number of workers and the time it takes to finish painting a house.

Number of workers	1	2	3	4	5	6
Time (hours)	10	5	3.3	2.5	2	1.7

   a. Are these two quantities in proportion?

   b. If so, write an equation relating the two and state the constant of proportionality.

13. The monthly cost for a cell phone depends on the total time the phone is used for calls, as shown in the graph.

   a. Are the two quantities, the cost C and the calling time t, in proportion?

   b. What are the coordinates of the marked point?

   c. What does that point mean in terms of this situation?

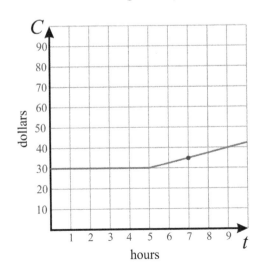

Cost of using a cell phone

Graphing Proportional Relationships – More Practice

You may use a calculator for all the problems in this lesson.

1. Shelly uses up a 400-ml bottle of shampoo in 7 months.

   a. What is the unit rate in this situation?

   b. Write an equation relating the amount of shampoo (S) in milliliters to the time (t) in months.

   c. Plot your equation. Choose appropriate scaling for the two axes.

   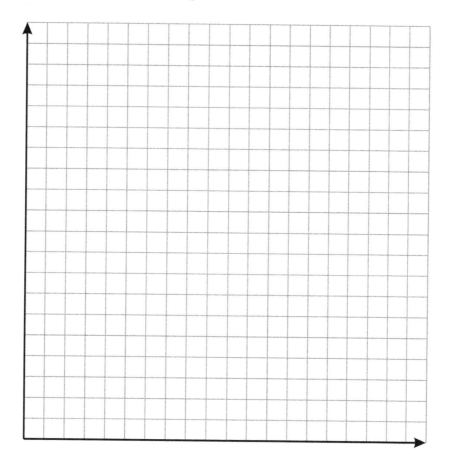

   d. Plot a point that corresponds to the time $t = 3$ months.

   e. Plot a point that corresponds to the unit rate.

   f. Using shampoo at the same rate, how long would it take her to use 250 ml of shampoo?

2. Sarah works as a secretary at a hospital. She gets paid $100 for an 8-hour workday.

   a. Write an equation that relates Sarah's pay to the time (in hours) that she has worked.

   b. Plot the equation on the grid. Choose the scaling on the axes so that you can fit the point that corresponds to *time* = 20 hours onto the grid.

   c. Plot the point that corresponds to the unit rate.

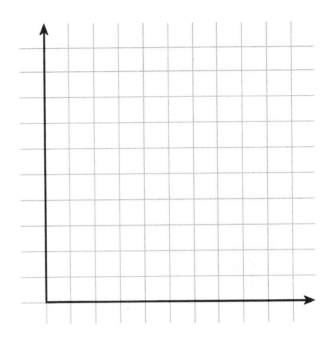

3. A car can travel 28 miles on a gallon of gasoline, and gasoline costs $3 a gallon. Notice that this situation actually involves *three* quantities: the mileage, the amount of gasoline used, and the cost of gasoline.

   a. We will now consider the mileage (*m*) and the amount of gasoline (*g*). Write an equation that gives you the mileage in terms of the gasoline used (in the form $m =$ (expression)).

   b. Plot your equation. Plot the independent variable on the horizontal axis. Make sure you can fit 20 gallons on the *g*-axis and the distance the car travels on 20 gallons of gasoline on the *m*-axis.

   c. Plot a point that corresponds (approximately) to $m = 100$ miles.

   d. Plot a point that corresponds to $g = 10$ gallons.

   e. Plot a point that corresponds to the unit rate.

   f. What does the point (0, 0) mean in this situation?

   g. How far could the car travel on 100 gallons of gasoline?

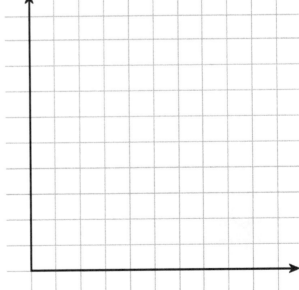

   Now we will also consider the third quantity, the *cost*.

   h. How far could the car travel on $100 of gasoline?

   i. Find the cost of traveling 700 miles.

36

More on Proportions

You can often solve a rate problem in several different ways. Try to find more than one way to solve the problems on this page. You can get some ideas from the example below.

Example 1. Tim can type 350 words in 9 minutes. Typing at the same speed, how many words could he type in 40 minutes?

Solution with a proportion:

$$\frac{350 \text{ words}}{9 \text{ min}} = \frac{w}{40 \text{ min}}$$

$$9 \text{ min} \cdot w = 350 \text{ words} \cdot 40 \text{ min}$$

$$\frac{9 \text{ min} \cdot w}{9 \text{ min}} = \frac{14{,}000 \text{ words} \cdot \text{min}}{9 \text{ min}}$$

$$w \approx 1556 \text{ words}$$

Solution with a unit rate:

The rate of 350 words per 9 minutes means $350 \div 9 = 38.\overline{8}$ words per minute.

In 40 minutes he can type 40 times that many words, or 40 min $\cdot$ 38.$\overline{8}$ words per minute = 1,555.$\overline{5}$ words ≈ 1556 words.

Solution in a diagram form:

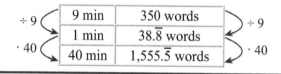

You may use a calculator for all the problems in this lesson. Choose one (or more) of the exercises 1-3 and try to find several ways to solve it.

1. The total weight of 18 identical books is 27.4 pounds.
   How much do five of those books weigh?
   Try to find more than one way to solve this problem.

2. Fencing costs $45 for a roll of 8 meters, but the vendor will also let you purchase part of a roll. So how much will it cost to fence a rectangular 15 m × 20 m plot?
   Try to find more than one way to solve this problem.

3. If a boy can ride his bike 15 km in 37 minutes, how far could he ride in one hour (at the same speed)?
   Try to find more than one way to solve this problem.

> **Comparing rates**
>
> Once again, there are several ways to determine if two rates are equal or which of two rates is greater. First try to solve the exercises 4 and 5 on your own! Then read the example at the bottom of the page.

4. One particular pasta sauce costs $3.95 for 450 g and another costs $4.55 for 560 g. Are the two rates equal? If not, which sauce costs more per gram?

5. Are these two rates equal? 50 miles / 2.2 gallons and 125 miles / 5.5 gallons. If not, which one is greater?

Example 2. John applied 10 kg of fertilizer to a field of 600 m², and the next day he applied 16 kg of fertilizer to a field of 1500 m². Are the rates equal?

(1) Solution using a proportion:	(2) Solution using the unit rates:
If the rates are equal, they are in proportion, and cross-multiplying will produce a true equation. $$\frac{10 \text{ kg}}{600 \text{ m}^2} \stackrel{?}{=} \frac{16 \text{ kg}}{1500 \text{ m}^2}$$ $$10 \text{ kg} \cdot 1500 \text{ m}^2 \stackrel{?}{=} 16 \text{ kg} \cdot 600 \text{ m}^2$$ $$15{,}000 \text{ kg} \cdot \text{m}^2 \neq 9{,}600 \text{ kg} \cdot \text{m}^2$$ We didn't get a true equation, so the rates are not equal.	The first unit rate is 10 kg/600 m² = $0.01\overline{6}$ kg/m². The second is 16 kg/1500 m² = $0.010\overline{6}$ kg/m². The unit rates are not equal, so the original rates are not equal either. The first unit rate (and thus the first rate) is actually quite a bit more than the second.

(3) Solution with logical reasoning

If 10 kg was enough for 600 m², and the next day he used 1.6 times that much (16 kg), then the area should be 1.6 times 600 m². However, 1500 m² is 2 1/2 times 600 m². So the rates are not equal.

6. Write a word problem that can be solved by the given proportion in the lower box. Then solve the proportion. Round your answer to a meaningful accuracy. Don't forget to check that your answer is reasonable.

a. Word problem:	b. Word problem:
a. Proportion: $$\frac{60 \text{ words}}{97 \text{ sec}} = \frac{10{,}000 \text{ words}}{x}$$	b. Proportion: $$\frac{\$5.59}{80.0 \text{ g}} = \frac{\$100}{x}$$

7. The cost of electricity to run an air conditioner for 24 hours is $3.60. How much would it cost to run the air conditioner for a full week if it is on only 14 hours a day?

8. A man can carry 90 lb. If ten identical books weigh 27 lb, then how many of those books can the man carry?

9. Elijah needs to apply fertilizer on the lawn at the golf club. The instructions say to apply 1 pound of nitrogen per each 1,000 square feet of lawn. The fertilizer he uses comes in 10-lb bags and is composed of 25% nitrogen and 75% other minerals. How many 10-lb bags of fertilizer does Elijah need for a rectangular lawn of 300 ft by 1,200 ft?

Puzzle Corner

Let's say that the ratio of quantities y and x is always 2/3. The table below shows some possible values of x and y.

x	1	2	3	4	5
y	1.5	3	4.5	6	7.5

Since the ratio x/y is always 2/3, we can write a proportion: $\dfrac{x}{y} = \dfrac{2}{3}$.

You have learned that if two quantities are in proportion, the equation relating them is in the form $y = mx$. Can the proportion above be written in the form of $y = mx$ for some constant m? If so, what is the value of m?

Scaling Figures

Two figures are **similar** if they have the same shape. Similar figures may be of different sizes. For example, all circles are similar, and so are all squares.

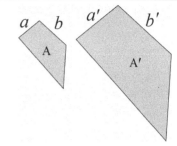

Example 1. The quadrilaterals A and A′ (read: "A prime") at the right are similar: they have the same basic shape, but one is larger.

Compare the corresponding sides: a to a' and b to b'. In the case of polygons, similarity means that **corresponding sides are proportional** (in the same ratio) and corresponding angles are equal.

So the ratio $a : a'$ is equal to the ratio $b : b'$. This ratio is called the **similarity ratio**.

Example 2. The similarity ratio between the two rectangles is 2:7. Find the length of the side marked x.

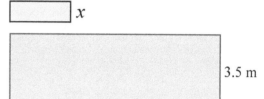

Solution 1. The lengths of the corresponding sides are in the ratio of 2:7. The unknown length of the side x corresponds to the 2 parts of the ratio and the known 3.5 m side corresponds to the 7 parts. So each part is 3.5 m ÷ 7 = 0.5 m. The unknown length is 2 · 0.5 m = 1 m.

That makes sense, since we would expect the side marked with x to be quite a bit shorter than 3.5 m.

Solution 2. We write the ratio of the lengths of the corresponding sides and set that ratio to be 2/7. We get an equation involving two equal ratios—a proportion. Its solution is on the right.

$$\frac{x}{3.5 \text{ m}} = \frac{2}{7}$$
$$7x = 2 \cdot 3.5 \text{ m}$$
$$7x = 7 \text{ m}$$
$$x = 1 \text{ m}$$

1. The figures are similar. Find the length of the side labeled x.

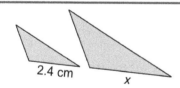

a. Similarity ratio 3:5.

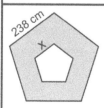

b. Similarity ratio 7:3.

2. The sides of two similar triangles are in a ratio of 3:4. If the sides of the larger triangle are 4.8 cm, 6.0 cm, and 3.6 cm, what are the sides of the smaller triangle?

3. Draw any triangle on blank paper or below. Then draw another, bigger triangle using the similarity ratio 2:5. Remember that corresponding angles in the two triangles will be equal.

4. The rectangles 1, 2, 3, and 4 in the table are similar.

   a. Consider the columns for length and width only. Complete the table by filling in the missing widths.

   b. In the last column, write the *aspect ratio* (the ratio of length to width) of each rectangle in simplified form. For example, for Rectangle 4, the aspect ratio is 2.5 cm : 7.5 cm = 25 : 75 = 1 : 3. What do you notice?

	Length	Width	Aspect Ratio
Rectangle 1	1 cm		
Rectangle 2	1.5 cm		
Rectangle 3	2 cm		
Rectangle 4	2.5 cm	7.5 cm	

Example 2. Sometimes you need to look carefully to find the corresponding sides. The two rectangles at the right are similar. Notice that the "top" sides of 18 ft and 22 ft do *not* correspond. Instead, the 18 ft side corresponds to the 10 ft side, because they are the *shorter* sides of the rectangles.

Here, the similarity ratio is 18 ft : 10 ft = 9:5. Also, since the side *x* units long corresponds to the 22 ft side, the ratio *x* : 22 ft is equal to ratio 9:5.

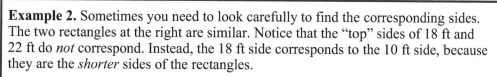

5. The figures are similar. Find the length of the side labeled with *x*.

a.
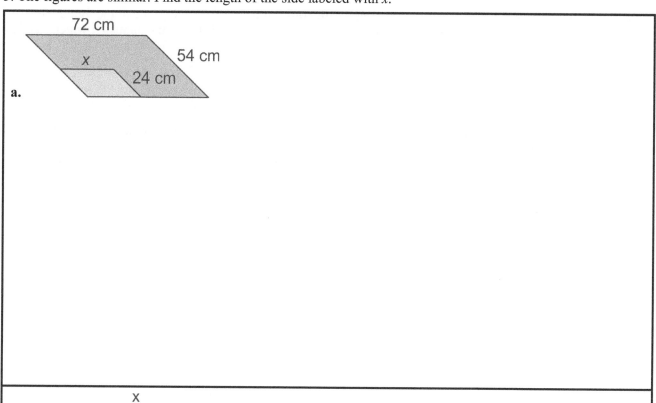

b.
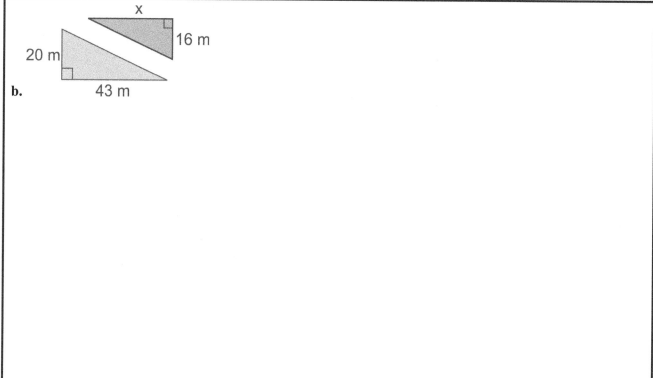

43

> **Scaling** means to enlarge or shrink a figure while maintaining its shape. The resulting figure is therefore similar to the original one. The number by which all distances or dimensions of the figure are multiplied is called the **scale factor** or just the **scale**.
>
> **Example 3.** A square was enlarged. What is the scale factor?
>
>
>
> Since the 48 cm-side became 72 cm long, the scale factor from the smaller square to the larger one is 72/48 = 3/2 = 1.5. So each side of the square became 1.5 times as long as before.
>
> 48 cm   72 cm
>
> We can also consider the **scale ratio** (or **ratio of magnification**): it is the ratio of any side of the figure after scaling it to its corresponding side before scaling (*after* : *before*). This ratio is the same, no matter which side is chosen. In this case, the scale ratio is 72:48 = 3:2. (Note that we don't write it as 48:72 because 48:72 is less than 1 and that would mean the figure shrank.)
>
> The scale factor and the scale ratio are equal. In this case, the scale factor was 3/2 or 1.5, and the scale ratio was 3:2. Simply put, you get the scale factor by writing the scale ratio as a fraction or as a decimal.
>
> Here is a way to keep the two very similar-sounding terms straight: The scale *ratio* is a ratio of two numbers (like 3:1), but the scale *factor* is a single number (such as 3).

6. **a.** Find the scale *factor* from the smaller to the larger parallelogram.

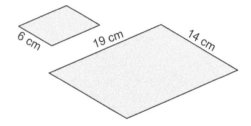

   b. What is the scale *ratio*?

7. A rectangle with 20 ft and 48 ft sides is shrunk so that its sides become 15 ft and 36 ft.

   a. What is the scale ratio?

   b. What is the scale factor?

8. The area of a square is 36 cm². The square is shrunk using the scale factor 3/4. What is the area of the resulting square?

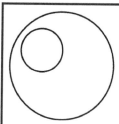

Example 4. A circle was shrunk. Find the scale ratio and the scale factor.

Any corresponding dimension — the diameter, the radius, or the circumference — will work for finding the scale ratio and scale factor. Measuring the diameters is the easiest.

The diameters measure 3.5 cm and 1.4 cm. Since the circle was shrunk, the scale ratio and the scale factor are less than 1, so we write the ratio as 1.4 : 3.5 and not vice versa.

However, it is customary to express ratios using whole numbers when possible. Multiplying both terms of the ratio by 10 gives us an equivalent ratio without decimals that we can reduce to lowest terms:

$$\frac{1.4}{3.5} = \frac{14}{35} = \frac{2}{5}$$

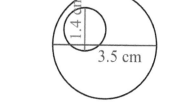

Writing 2/5 as a decimal, we get the scale factor 0.4. So the dimensions of the smaller circle are 2/5 or 0.4 times the dimensions of the larger one.

9. The triangle on the left was scaled to become the triangle on the right. Find the scale ratio, and then write it as a scale factor. Use a ruler that measures in centimeters.

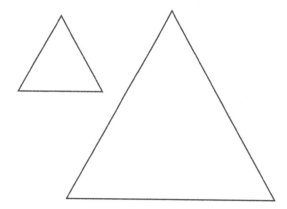

10. Enlarge this L-shape using the scale ratio 3:2. Draw the resulting larger L-shape beside it.

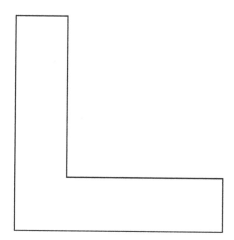

11. The sides of a rectangle measure 3" and 4 1/2". The shorter side of another, similar rectangle is 3/4".

   a. In what ratio are the sides of the two rectangles?

   b. What is the length of the other side of the similar rectangle?

   c. Calculate the areas of both rectangles.

   d. In what ratio are their areas?

Puzzle Corner

The aspect ratio of a rectangle is 2:3 and its perimeter is 50 cm. The rectangle is shrunk in a scale ratio of 2:5. What is its area now?

Floor Plans

Floor plans are drawn using a **scale**, which is a ratio relating the distances in the plan to the distances in reality. For example, a scale of 1 cm : 2 m means that 1 cm in the drawing corresponds to 2 m in actual size.

Example 1. In reality, how big is a room that measures 1 ¾" by 2 ½" in a plan with a scale of 1 in : 10 ft?

Since 1 inch corresponds to 10 ft, we simply need to multiply the length and width given in inches by 10.

Using decimals, the dimensions are 1.75 in by 2.5 in. So the dimensions of the room in actual size are 1.75 · 10 = 17.5 ft and 2.5 · 10 = 25 ft.

However, we're really not just multiplying by the number 10 but by the ratio 10 ft/1 in. That's how we keep track of the units to make sure that our final answer ends up with the correct units (feet and not inches). This is what's really happening in the calculation:

$$1.75 \text{ in.} \cdot \frac{10 \text{ ft}}{1 \text{ in.}} = 17.5 \text{ ft} \quad \text{and} \quad 2.5 \text{ in.} \cdot \frac{10 \text{ ft}}{1 \text{ in.}} = 25 \text{ ft}$$

Why not multiply by 1 in/10 ft (or 1 in : 10 ft) as the ratio is stated in the problem? Then the inches ("in") in the dimension wouldn't cancel the inches in the conversion factor, and we would end up "in²/ft" as our unit of length, instead of "ft."

Example 2.

A room measures 4.2 m by 3.2 m. How big would it appear on a floor plan with a scale 1 cm : 0.8 m?

Instead of multiplying by the scale ratio like we did in Example 1, we can set up one proportion for the length (L) and another for the width (W) and solve them. The length is 5.25 cm, and the width is 4 cm.

$$\frac{1 \text{ cm}}{0.8 \text{ m}} = \frac{L}{4.2 \text{ m}} \qquad \frac{1 \text{ cm}}{0.8 \text{ m}} = \frac{W}{3.2 \text{ m}}$$

$$0.8L = 4.2 \text{ cm} \qquad 0.8W = 3.2 \text{ cm}$$

$$\frac{0.8L}{0.8} = \frac{4.2 \text{ cm}}{0.8} \qquad \frac{0.8W}{0.8} = \frac{3.2 \text{ cm}}{0.8}$$

$$L = 5.25 \text{ cm} \qquad W = 4 \text{ cm}$$

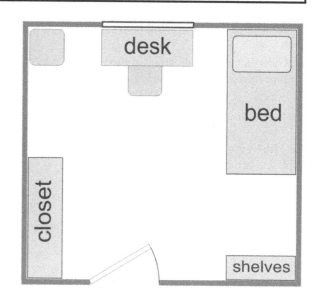

1 in : 4 ft

1. This room is drawn at a scale of 1 in : 4 ft. Measure dimensions asked below from the picture and then calculate the actual (real) dimensions.

   a. the bed

   b. the desk

2. What is the area of this room in reality?

3. In the middle of the plan for the room, draw a table that in reality measure 3.5 ft × 2.5 ft.

47

4. A room measures 4 ½ inches by 4 inches in a plan with a scale of 1 in : 3 ft.

   a. What would the dimensions of the room be if it was drawn to the scale of 1 in : 6 ft?

   b. What would the dimensions of the room be if it was drawn to the scale of 1 in : 4 ft?

5. This room measures 2.5 m by 3.5 m in reality.

   a. To what scale is it drawn here?

   b. On the plan, to scale, draw:

   - two windows that are 80 cm wide,

   - a door that is 1 m wide, and

   - a bed that is 150 cm by 200 cm.

6. A floor plan is drawn using the scale 5 cm : 1 m.

   a. Calculate the dimensions in the plan for a kitchen that measures 4.5 m by 3.8 m in reality.

   b. The living room measures 26 cm by 22.5 cm on the plan. What are its dimensions in reality?

7. In the space at the right, draw a plan for a room that measures 3.8 m by 4.6 m. Put a 120 cm by 120 cm table in the middle of the room. Use the scale 2 cm : 1 m.

8. This is a floor plan for a small cottage at the scale 1/8 in : 1 ft. We have omitted the doors and windows to keep it very simple. (You can add some if you would like.)

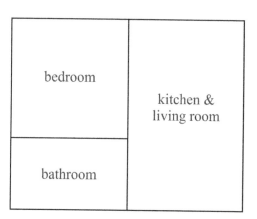

Scale 1/8 in : 1 ft

a. What is the area of the cottage in reality?

b. Redraw the plan at the scale 3/8 in : 1 ft on blank paper.

9. Redraw this floor plan at the scale 1 cm : 125 cm.

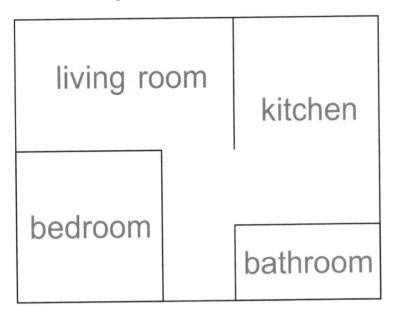

Scale 1 cm : 1 m

Maps

Just like floor plans, maps also include a scale. A scale on a map may show how units on the map correspond to units in reality (for example 1 cm = 50 km). It can also be given as a ratio such as 1:120,000.

A scale of 1:120,000 means that 1 unit on the map corresponds to 120,000 units in reality. This holds true—whether you use centimeters, millimeters, or inches—because the scale 1:120,000 is a ratio without any particular unit. So 1 cm on the map corresponds to 120,000 cm in reality, and 1 inch on the map corresponds to 120,000 inches in reality.

Example 1. A map has a scale 1:150,000. How long in reality is a distance of 7.1 cm on the map?

Below you can read two solutions to this problem. Both are actually very similar!

Multiply, then change the units.

If 1 cm corresponds to 150,000 cm, then 7.1 cm corresponds to 7.1 · 150,000 cm = 1,065,000 cm.

To be useful, this figure needs to be converted into kilometers. You can do this in two steps:

1. From centimeters to meters: Since 1 m = 100 cm, we remove two zeros from 1,065,000 cm to get 10,650 meters (or you can think of it as dividing by 100).
2. From meters to kilometers: Since 1 km = 1,000 m, the 10,650 meters corresponds to 10.65 km ≈ 11 km.

Change the units, then multiply.

In this solution, we will first rewrite the scale and then use multiplication to calculate the distance in reality.

Since 1 cm corresponds to 150,000 cm, and 150,000 cm = 1,500 m = 1.5 km, we can rewrite the scale of this map as 1 cm = 1.5 km.

Then, 7.1 cm corresponds to 7.1 · 1.5 km = 10.65 km ≈ 11 km.

You can use a calculator for all the problems in this lesson.

1. A map has a scale ratio of 1:20,000. Fill in the table.

on map (cm)	in reality (cm)	in reality (m)	in reality (km)
1 cm	20,000 cm		
3 cm			
5.2 cm			
0.8 cm			
17.1 cm			

2. A map has a scale of 1:100,000.

   a. The scale says that 1 cm on the map corresponds to 100,000 cm in reality. How many kilometers is that?

   Thus, we can rewrite this scale in the format 1 cm = _____ km

   b. A ski trail measures 5.2 cm on this map. In reality, how long is the trail in kilometers?

3. A map has a scale of 1:25,000.

   a. Rewrite this scale in the format 1 cm = _____ m.

   b. Fill in the table. Give your answers to the nearest tenth of a centimeter.

on the map (cm)	in reality
	500 m
	900 m
	1.6 km
	2.5 km

4. Measure the aerial distances between the given places in centimeters and then calculate the distances in reality to the nearest kilometer. The places are marked with squares on the map. *Aerial distances* are "as the crow flies": measure them directly from point to point, not by following the roads.

   a. From Elkmont to the the Gatlinburg Welcome Center.

   b. From the Great Smoky Mountains Institute at Tremont to the Little Greenbrier School.

   c. From the Little Greenbrier School to Elkmont.

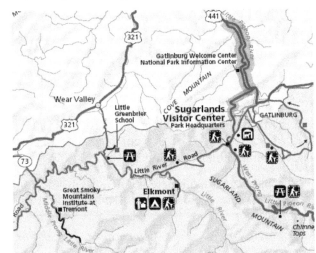

Scale 1:180,000

5. Hannah is making a map of a farmhouse and the surrounding buildings. When she measured the distance from the farmhouse to the barn, it was 75 meters.

   a. On the map, what will the distance be from the house to the barn at a scale of 1:500?

   b. What would the distance be on a map at a scale of 1:1200?

Example 2. The distance from Jane's home to her grandma's is 220 km.
How long is the representation of that distance on a map with a scale of 1:1,500,000?

Study both solutions below, and make sure you understand them.

Divide, then convert.	**Convert, then divide.**
This conversion goes the other way: from reality to the map. Therefore, we need to *divide* the distance 220 km by the factor 1,500,000. We will get a very small number, and it is in kilometers just like the original distance is: $$220 \text{ km} \div 1{,}500{,}000 = 0.00014\overline{6} \text{ km}$$ However, the answer in this format is not very useful. We need to convert it into units that can be measured on a map, such as centimeters (millimeters would work, too). You could convert $0.000014\overline{6}$ km into centimeters directly, but here we will do it in two steps because that is easier for most people. (1) Converting from kilometers to meters requires multiplying our number by the unit ratio 1,000 m/km (not by 1 km/1000 m because then the units "km" wouldn't cancel): $$0.000146\overline{6} \text{ km} \cdot \frac{1{,}000 \text{ m}}{\text{km}} = 0.14\overline{6} \text{ m}$$ (2) Finally, we convert meters into centimeters by multiplying by 100 cm/m. We get $$0.14\overline{6} \text{ m} \cdot \frac{100 \text{ cm}}{\text{m}} = 14.\overline{6} \text{ cm}.$$ So the distance on the map is about 14.7 cm.	We will first rewrite the scale and then use division to calculate the distance in reality. According to the scale, 1 cm corresponds to 1,500,000 cm. Now, 1,500,000 cm = 15,000 m (think of dropping two zeros) and that equals 15 km (think of dropping three zeros). So we can rewrite the scale of this map as 1 cm = 15 km. Thus 220 km corresponds to 220 km/15 km = $14.\overline{6} \approx 14.7$ cm.

6. Mark is planning a route for a footrace that will be 1.5 km long. He has two city maps available. One has a scale of 1:15,000 and the other has a scale of 1:20,000. Calculate the distance of the race on each of the two maps to the nearest tenth of a centimeter.

Example 3. A map has a scale of 1:500,000. The distance from one town to another measures 5 1/4 inches on the map. How long is the distance in reality?

Again, there are two ways to solve this problem:

Multiply, then convert: Multiply the given distance 5.25 in by 500,000 and then convert the result into miles.

Convert, then multiply: Rewrite the scale into an "easier" format, then use multiplication.

Multiply, then convert.

We simply multiply 5.25 in by 500,000 to get 5.25 in · 500,000 = 2,625,000 inches. So that is the distance in reality. Next we convert this distance into miles in two steps:

(1) From inches to feet: Since 1 ft = 12 in, we multiply 2,625,000 inches by the ratio 1 ft/12 in (not by 12 in/1ft because we want the inches to cancel):

$$2{,}625{,}000 \text{ in} \cdot \frac{1 \text{ ft}}{12 \text{ in}} = 218{,}750 \text{ ft}$$

You end up dividing 2,625,000 by 12, which makes sense since we are going from smaller units (inches) to bigger ones (feet), and thus we should get *fewer* units in feet.

(2) From feet to miles: 1 mi = 5,280 ft. Again, we multiply by the conversion ratio 1 mi/5,280 ft:

$$218{,}750 \text{ ft} \cdot \frac{1 \text{ mi}}{5{,}280 \text{ ft}} = 41.4299\overline{24} \text{ mi} \approx 41 \text{ miles}.$$

(Essentially, you divide by 5,280.)

The distance between the towns is about 41 miles.

Convert, then multiply.

The scale 1:500,000 means that 1 inch corresponds to 500,000 inches. We need to convert that to a more useful unit, such as feet or miles. Similar to above, the conversions go like this:

$$500{,}000 \text{ in} \cdot \frac{1 \text{ ft}}{12 \text{ in}} = 41{,}666.\overline{6} \text{ ft} \quad \text{and} \quad 41{,}666.\overline{6} \text{ ft} \cdot \frac{1 \text{ mi}}{5{,}280 \text{ ft}} = 7.89\overline{14} \text{ mi} \approx 7.90 \text{ miles}$$

So the scale of our map is 1 inch = 7.90 miles.

Then, the given distance 5.25 miles corresponds to 5.25 · 7.90 miles ≈ 41 miles.

Either way, the most difficult part is in converting inches into miles. Since we are dealing with customary units, there aren't any shortcuts that would allow us to convert the measuring units without a calculator, so neither solution is really easier than the other.

---

7. A map has a scale ratio of 1:400,000. In miles, how long is a nature hike that measures 2.5 inches on the map? Give your answer to the nearest mile.

8. Use a map you have on hand, and measure distances on it with a ruler. Then calculate the distances in reality and give them to a reasonable accuracy. If you don't have a map on hand, skip this exercise and just do the next one.

9. On this map of the USA measure the distances in inches. Then calculate the distances in reality.

   a. From Tallahassee to Denver.
      (Round to the nearest hundred miles.)

   b. From Sacramento to Austin.
      (Round to the nearest hundred miles.)

   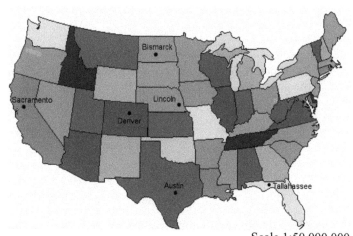

   Scale 1:50,000,000

   c. From Lincoln to Bismarck.
      (Round to the nearest ten miles.)

10. An island is 16.2 miles from the mainland. What is that distance on a map with a scale of 1:500,000? Finish Ellie's solution to this problem by filling in the words *multiply* and *divide*, the sign "·" or "÷," and numbers. Give your final answer to the tenth of a inch.

    First, I _____ the distance 16.2 miles by the factor 500,000. I will get a very small number, which will be in miles: 16.2 miles ▭ 500,000 = _____ miles.

    Next I convert this to feet, and then to inches.

    Converting miles to feet means to _____ by the ratio 5,280 ft/1 mi:

    $$_____ \cdot \frac{5{,}280 \text{ ft}}{1 \text{ mi}} = _____$$

    Then I convert the result from feet to inches by _____ing by the ratio 12 in/ 1 ft:

    $$_____ \cdot \frac{12 \text{ in.}}{1 \text{ ft}} = _____ \text{ in.} \approx _____ \text{ in.}$$

11. The distance from Mark's home to the airport is 45.62 miles according to an online distance calculator. How long would this distance be, in inches, on a map with a scale of 1:250,000? How about on a map with a scale of 1:300,000?

12. The scale of a map is 1:15,000. A rectangular plot of land measures 1 3/16″ by 2 1/8″ on the map.
    a. Find the area of the land in reality in square feet. Don't round your answer, as we will use the answer in part (b).

    b. Calculate the area of the land in acres, to the nearest tenth of an acre.
    Use 1 acre = 43,560 square feet.

13. The length of a hiking path is 5.0 inches on a map with a scale of 1:200,000. What would the length be on a map with a scale of 1:150,000?

Puzzle Corner

A sheet of A4 paper measures 210 mm by 297 mm. You want to print a map of a plot of land with the dimensions of 1.65 km by 2.42 km onto one sheet of A4 paper. What scale should you use for your map so that it fits onto the sheet of A4 paper?

Significant Digits

(This lesson is optional.)

Example 1. In reality, how long is a distance of 5.7 cm on a map with a scale of 1:400,000?

Since 1 cm corresponds to 400,000 cm, then 5.7 cm corresponds to 5.7 · 400,000 cm = 2,280,000 cm. Converting this into kilometers we get 22.8 km.

However, since our measurement was only to the accuracy of a tenth of a centimeter, we cannot truthfully give our answer to an accuracy of 22.8 km. You see, the measurement 5.7 cm is an *approximation*. The true distance on the map could be 5.7352 cm or 5.67364 cm — we don't know since we cannot measure it that accurately.

Let's consider some other distances on the map that would be rounded to 5.7 cm, and calculate them in reality. Study the table on the right:

From the table we can see that the distance in reality is anywhere from 22.6 km to about 23 km. We definitely cannot say it is exactly 22.8 km. That is why we need to round 22.8 km *to the nearest kilometer*. The distance in reality is about 23 km.

5.65 cm on map = 22.6 km in reality
5.688 cm on map = 22.752 km in reality
5.703 cm on map = 22.812 km in reality
5.718 cm on map = 22.872 km in reality
5.749 cm on map = 22.996 km in reality

Significant digits of a number are those digits whose value contributes to the precision of the number. Significant digits help us know how to round answers when calculating *measurements*, because measurements by their nature are never totally precise.

For example, all the individual digits of 12.593 m tell us something about its precision: it is precise to the thousandth of a meter. However, in the measurement 2,000 m, we cannot be sure if the number was originally measured as 1,9283.4 m and rounded to 2,000 m or measured as 2,400 m and rounded to 2,000 m. So in 2,000 m, only the 2 is a significant digit that tells us something about its precision.

All non-zero digits are always significant. With zeros, the situation is more complex. Here are the rules:

1. All non-zero digits are significant: 38.2 has three significant digits.
2. Zeros between other significant digits are also significant: 50,039 has five significant digits.
3. Non-decimal zeros at the end of a number are not significant: 6,400 has two significant digits.
4. Decimal zeros in front of the number are not significant: 0.0038 has two significant digits.
5. Decimal zeros at the end of a number *are* significant: 0.00380 has three significant digits.

In a calculation involving multiplication and/or division, the amount of significant digits in the answer should equal the amount of significant digits in the number that is the least precise (that has the smallest amount of significant digits).

For example, 2.3 cm has two significant digits and 11.9 cm has three. When we multiply them (to get an area), we get 27.37 cm^2, but we need to take the result to only *two* significant digits (because 2.3 cm had the least amount of significant digits, which was 2), so 27.37 cm^2 gets rounded to 27 cm^2.

In this lesson you will often multiply or divide a measurement result by a conversion factor. In this situation, **keep the same number of significant digits in your converted result as what you had in your measurement.** That is because the conversion factors are more exact and have more significant digits than your measurement result, so the measurement will automatically be the number with the least amount of significant digits.

You can use a calculator for all the problems in this lesson.

1. How many significant digits do these numbers have?

a. 24.5 km	b. 20.5 km	c. 24.50 km	d. 0.5 mi
e. 15,000 ft	f. 15,001 ft	g. 0.078 km	h. 0.0780 km
i. 5,002.90 kg	j. 340 lb	k. 340.9 lb	l. 0.005 lb

2. The two sides of a rectangular play area are measured to be 24.5 m and 13.8 m.

   a. Calculate its area and give it with a reasonable amount of significant digits.

   b. Let's say the dimensions of the play area were measured more accurately to be 24.56 m and 13.89 m. Calculate the area and give the result to a reasonable accuracy.

3. Calculate the following distances in reality. Consider how many significant digits your answer should have. Note: All digits in the scale ratios are significant. For example, the scale ratio of 1:50,000 is precise to all 5 digits. It's neither 1:49,999 nor 1:50,001, but exactly 1:50,000.

   a. 6.2 cm on a map with a scale of 1:50,000

   b. 12.5 cm on a map with a scale of 1:200,000

   c. 0.8 cm on a map with a scale of 1:15,000

4. A field measures 5.0 cm by 3.5 cm on a map with a scale of 1:8,000.
   Calculate its area in reality.

5. The distance from Mary's home to school is 3.0 inches on a map with a scale of 1:10,000.

   a. How long is this distance in reality? Give your answer in miles to two significant digits.

   b. Give your answer in yards to two significant digits.

6. A gas station is on a rectangular plot of land that measures 45.0 m by 31.2 m.
   What are these dimensions on a map with a scale of 1:500?

58

Chapter 6 Mixed Review

1. The table lists the average high temperatures for January and July for several Canadian cities. It gives you an idea of how warm it gets during the winter and during the summer, on average.

City	January (Avg. High °C)	July (Avg. High °C)
Winnipeg, MB	−11.9	25.9
Saskatoon, SK	−10.1	25.3
Quebec City, QC	−7.0	24.7
Edmonton, AB	−6.3	22.8
Ottawa, ON	−5.8	26.6
Calgary, AB	−0.9	23.2
Montreal, QC	−5.3	26.3
Halifax, NS	−0.1	23.1
Toronto, ON	−0.7	26.6
Vancouver, BC	6.8	22.1
Yellowknife, NT	−21.6	21.3
Iqaluit, NU	−22.8	12.3

   a. In which city is the difference between the average high temperatures in these two months the greatest?

   b. How much is that difference?

   c. In which city is the difference the smallest?

   d. How much is that difference?

2. Write the numbers in scientific notation.

   a. 2,089,000

   b. 394,410,000

3. Divide and simplify.

a. $2 \div (-8) = -\dfrac{2}{8} = -\dfrac{1}{4}$	**b.** $4 \div (-24)$	**c.** $-18 \div 5$
d. $-2 \div (-9)$	**e.** $42 \div (-49)$	**f.** $-32 \div (-28)$

4. Find the value of the expressions using the correct order of operations.

a. $5 \cdot \dfrac{2}{-10}$	**b.** $-\dfrac{12}{-4} + 7$	**c.** $-1 + \dfrac{24}{12 + (-6)}$
d. $-2 + 7 \cdot 2 - 6$	**e.** $-8 \cdot (-7) - 11$	**f.** $(-3 + 9) \cdot 8$

5. Solve. Check your solutions (as always!).

a. $11 - 5x = -6$

Check:

b. $6(y + 2) = -16$

Check:

c. $\dfrac{2x}{5} = 30$

Check:

d. $\dfrac{s - 12}{5} = -1$

Check:

6. Find the slope of the line.

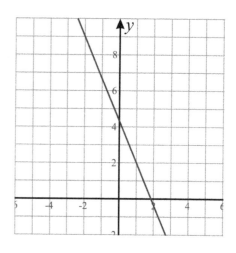

7. Write an inequality for each situation. Use a variable for the quantity in question (the temperature, the amount of flour, and the cost).

   a. The temperature shouldn't exceed 42°C.

   b. I need at least 3 cups of flour.

   c. The cost has to be kept strictly under $3,000.

8. A farmer is hiring workers to help him on his farm. He will pay each worker a monthly salary of $2,050. The farmer has a total budget of $40,000 for three months. How many workers can he hire?

   a. Solve the problem without using an equation or inequality.

   b. Write an inequality for the problem, and solve it.

9. **a.** Plot the equation $y = (3/4)x$.

   b. Plot the equation $y = -2x + 1$.

   c. Plot the equation $y = 4 - x$.

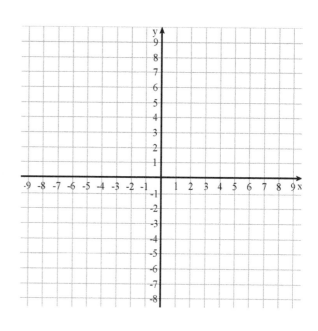

10. Determine whether the point $(2, -3)$ is on the line $y = -2x - 2$. Justify your answer.

Chapter 6 Review

1. Simplify the ratios and rates.

a. 164 km per 4 hours	b. $\dfrac{6 \text{ g}}{1600 \text{ ml}} =$	c. 52 : 156 = _____ : _____

2. A car traveled 348 miles in 6 hours. Fill in the table of equivalent rates.

Miles						348		
Hours	1	2	3	4	5	6	10	20

3. A mixture of salt and water contains 20 grams of salt and 1,200 grams of water.
   Write the ratio by weight of salt to water and simplify it.

4. Susan can jog 1 1/2 miles in 1/3 hour.
   Write a rate for her jogging speed and simplify it.

5. Solve the proportions. Round your answers to the nearest hundredth.

a. $\dfrac{16}{17} = \dfrac{109}{T}$	b. $\dfrac{1.5}{2.8} = \dfrac{M}{5}$

6. Write a proportion for the following problem and solve it.

   12 kg of chicken feed costs $19.
   How much would 5 kg cost?   _____ = _____

7. On the average, Gary makes a basket eight times out of every ten shots. How many baskets can he expect to make when he practices 25 shots?

8. Write the unit rate as a complex fraction, and then simplify it.

a. Alex solved 2 1/2 pages of math problems in 1 1/4 hour.
b. Noah painted 2/3 of a room in 3/4 of an hour.

9. A car is traveling at a constant speed of 75 km per hour.

   a. Write an equation relating the distance (d) and time (t) and plot it in the grid below.

   b. What is the unit rate?

   c. Plot the point that matches the unit rate in this situation.

   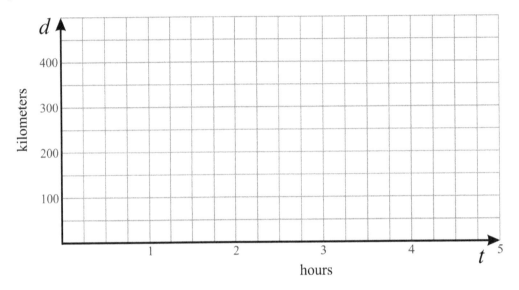

   d. What does the point (0, 0) mean in terms of this situation?

   e. How far can the car travel in 55 minutes, driving at the same speed?
   Also, plot the point for the time $t = 55$ min.

   f. How long will the car take to travel 160 km? Give your answer in hours and minutes.
   Also, plot the point that matches the distance $d = 160$ km.

10. Using a pre-paid internet service you get a certain amount of bandwidth to use for the amount you pay. The table shows the prices for certain amounts of bandwidth.

Bandwidth	1G	2G	5G	10G	15G	20G	25G
Price	$10	$16	$23	$30	$37	$43	$50

a. Are these two quantities in proportion?

Explain how you can tell that.

b. If so, write an equation relating the two and state the constant of proportionality.

11. In the year 2008 it was estimated that it cost $9,369 a year to drive a medium-sized car (a sedan) for 15,000 miles (a typical amount of use). Based on those same assumptions, how much would it cost, to the nearest dollar, to drive that car for 5 months?

12. The figures are similar. Find the length of the side labeled with x.

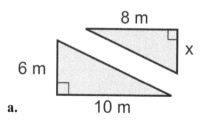

a.

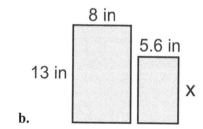

b.

65

13. A house plan has a scale of 1 in : 6 ft. In the plan, one room measures 2 in × 2 ¾ in. What are the true dimensions of the room?

14. A freight truck fully loaded with cargo gets six miles to a gallon of diesel.

   a. What is the unit rate in this situation?

   b. Write an equation relating the mileage (M) to the amount of diesel fuel (f) in gallons.

   c. Plot your equation. Choose an appropriate scaling for the two axes.

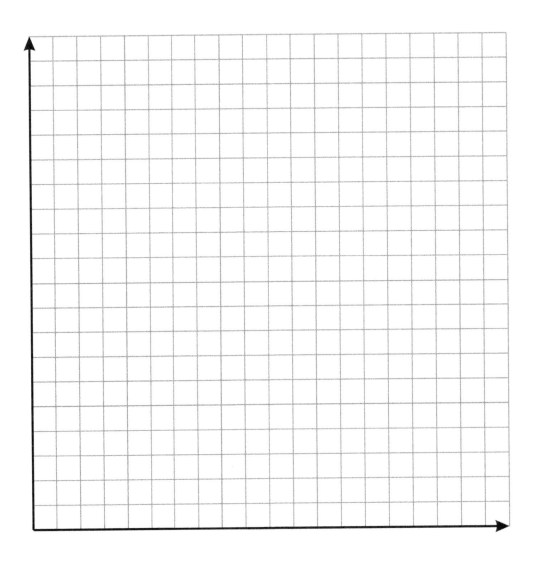

Chapter 7: Percent
Introduction

In this chapter we review the concept of percent as "per hundred" or as hundredth parts and how to convert between fractions, decimals, and percents. The lesson *Solving Basic Percentage Problems* is intended for review of sixth grade topics, focusing on finding a known percentage of a number (such as 21% of 56) or finding a percentage when you know the part and the total.

We take a little different perspective of these concepts in the lesson *Percent Equations*. Students write simple equations for situations where a price increases or decreases (discounts). This lesson also explains what a percent proportion is. Personally, I prefer *not* to use percent proportion but to write the percentage as a decimal and then write an equation. I feel that approach adapts better to solving complex problems than using percent proportion.

Here is a quick example to show the difference between the two methods. Let's say an item is discounted by 22% and it now costs $28. Then, the new price is 78% of the original. If we let p be the price of the item before the discount, we can write the percent proportion $28/p = 78/100$ and solve for p. If, we write the percentage 78% as the decimal 0.78, we get the equation $0.78p = \$28$. Personally, I consider percent proportion to be an optional topic, and the reason I have included it here is to make this curriculum fully meet the Common Core Standards for seventh grade.

The lesson *Circle Graphs* provides students a break from new concepts and allows them to apply the concept of percent in a somewhat familiar context. Next, we delve into the percentage of change. Students sometimes view the percentage of change as a totally different concept as compared to other percentage topics, but it is not that at all. To calculate the percentage of change, we still use the fundamental idea of *percentage = part/total*, only this time, the "part" is how much the quantity in question changes (the difference) and the "total" is the original quantity.

Tying in with percentage of change, students also learn to compare values using percentages, such as how many percent more or less one thing is than another. Once again, this is not really a new concept but is based on the familiar formula *percentage = part/total*. The percentage difference (or relative difference) is the fraction (*actual difference*)/(*reference value*).

Simple Interest is a lesson on the important topic of interest, using as a context both loans and savings accounts. Students learn to use the formula $I = prt$ in a great variety of problems and situations.

The text concludes with a review lesson of all of the concepts taught in the chapter.

You can find matching videos for topics in this chapter at http://www.mathmammoth.com/videos/ (grade 7).

The Lessons in Chapter 7

	page	span
Review: Percent	72	*3 pages*
Solving Basic Percentage Problems	75	*3 pages*
Percent Equations	78	*5 pages*
Circle Graphs	83	*2 pages*
Percentage of Change	85	*3 pages*
Percentage of Change: Applications	88	*4 pages*
Comparing Values Using Percentages	92	*4 pages*
Simple Interest	96	*6 pages*
Chapter 7 Mixed Review	102	*3 pages*
Chapter 7 Review	105	*2 pages*

Helpful Resources on the Internet

Percent Videos by Maria
Videos on percent-related topics that match the lessons in this chapter.
http://www.mathmammoth.com/videos/prealgebra/pre-algebra-videos.php#percent

Percent Worksheets
Create an unlimited number of free customizable percent worksheets to print.
http://www.homeschoolmath.net/worksheets/percent-decimal.php
http://www.homeschoolmath.net/worksheets/percent-of-number.php
http://www.homeschoolmath.net/worksheets/percentages-words.php

PERCENTAGES, FRACTIONS, AND DECIMALS

Matching Fractions and Percent
Practice matching percentages with corresponding fractions in this timed matching game.
https://www.mathplayground.com/matching_fraction_percent.html

Decention Game
Build teams of three players by matching fractions, decimals, and percentages with the same value.
https://www.mathplayground.com/Decention/index.html

Fractions and Percent Matching Game
A simple matching game: match fractions and percentages.
http://www.mathplayground.com/matching_fraction_percent.html

What percentage is shaded?
Practice guessing what percentage of the pie chart has been shaded yellow in this interactive activity.
http://www.interactivestuff.org/sums4fun/pietest.html

Pie Chart and Questions
First, read a short illustrated lesson about pie charts. Then, click on the questions at the bottom of the page to practice.
https://www.mathsisfun.com/data/pie-charts.html

Flower Power
Grow flowers and harvest them to make money in this addictive order-'em-up game. Practice ordering decimals, fractions, and percentages.
https://www.mangahigh.com/en/games/flowerpower

Matching Fractions, Decimals, and Percentages
A simple matching memory game.
http://nrich.maths.org/1249

Doughnut Percents
This task involving the equivalence between fractions, percentages and decimals is one of a series of problems designed to develop students' team working skills.
https://nrich.maths.org/6945

Percent Goodies: Fraction-Decimal-Percent Conversions
Practice conversions between fractions, decimals and percents. There are three levels of difficulty and instant scoring for each. Note that fractions must be written in lowest terms.
http://www.mathgoodies.com/games/conversions/

BASIC PERCENTAGE CALCULATIONS

Penguin Waiter
A simple game where you calculate the correct tip to leave the waiter (levels "easy" and "medium"), the percentage that the given tip is (level "hard"), or the original bill (level "Super Brain").
https://www.funbrain.com/games/penguin-waiter

Percent Jeopardy
An interactive jeopardy game where the questions have to do with a percentage of a quantity.
http://www.quia.com/cb/42534.html

Matching Percentage of a Number
Match cards that ask for a percentage of a number (such as 75% of 40) with the values. The game is fairly easy and can be completed using mental math.
http://www.sheppardsoftware.com/mathgames/percentage/MatchingPercentNumber.htm

Percent, Interest, Discount, and Sale Price Challenge Board
Score points by answering questions correctly. This game can be played by one or two players.
https://www.quia.com/cb/55701.html

The Percentage Game
This is a printable board game for 2-3 players that practices questions such as
"20 percent of ___ is 18" or "___ is 40 percent of 45".
http://nzmaths.co.nz/resource/percentage-game

Percentages
This page gives an illustrated explanation of the basic concept of percent. At the bottom of the page, there is also a series of practice questions.
https://www.mathsisfun.com/percentage.html

PERCENT OF CHANGE

Percent of Change Matching
Match five flashcards with given increases or decreases (such as "25 is decreased to 18") with five percentages of increase/decrease.
https://www.studystack.com/matching-182854

Percent Shopping
Choose toys to purchase. In level 1, you find the sale price when the original price and percent discount are known. In level 2, you find the percent discount (percent of change) when the original price and the sale price are known.
http://www.mathplayground.com/percent_shopping.html

Rags to Riches: Percent Increase or Decrease
Answer simple questions about percent increase or decrease and see if you can win the grand prize in the game.
http://www.quia.com/rr/230204.html

Percentage Change 1
A self-marking quiz with 10 questions about percentage change. The link below goes to level 1 quiz, and at the bottom of that page you will find links to level 2, 3, 4, 5 and 6 quizzes.
http://www.transum.org/software/SW/Starter_of_the_day/Students/PercentageChange.asp

Percent of Change Quiz
Practice determining the percent of change in this interactive multiple-choice quiz.
http://www.phschool.com/webcodes10/index.cfm?wcprefix=bja&wcsuffix=0607&area=view

Percentage Increase and Decrease 4 in a Line
The web page provides a game board to print. Players take turns picking a number from the left column, and increase or decrease it by a percentage from the right column. They cover the answer on the grid with a counter. The first player to get four counters in a line wins.
https://www.tes.co.uk/teaching-resource/percentage-increase-and-decrease-4-in-a-line-6256320

Percentage of Increase Exercises
Find the percentage increase given the original and final values in this self-check quiz about percentage change.
http://www.transum.org/software/SW/Starter_of_the_day/Students/PercentageChange.asp?Level=3

Percentage of Decrease Exercises
Find the percentage decrease given the original and final values in this self-check quiz about percentage change.
http://www.transum.org/software/SW/Starter_of_the_day/Students/PercentageChange.asp?Level=4

Treasure Hunt - Percentage Increase and Decrease
The clues of this treasure hunt are printable percentage increase/decrease questions.
https://www.tes.co.uk/teaching-resource/treasure-hunt--percentage-increase-and-decrease-6113809

Percent Change Practice
Interactive flash cards with simple questions about percentage of change with three difficulty levels.
http://www.thegreatmartinicompany.com/percent-percentage/percent-change.html

Percent of Change Test
Multiple-choice questions about percentage of change to be solved without a calculator (mental math).
https://reviewgamezone.com/mc/candidate/test/?test_id=29446&title=Percent%20Change

Percent of Change Jeopardy
This is an online jeopardy game that provides you the game board, questions for percent increase, percent decrease, sales tax, discounts, and markups, the answers, and a scoreboard where you can enter the teams' points. However, it doesn't have a place to enter answers and requires someone to supervise the play and the teams' answers.
http://www.superteachertools.us/jeopardyx/jeopardy-review-game.php?gamefile=2245685

Percentage Difference
A short lesson about percentage difference, followed by practice questions at the bottom of the page.
http://www.mathsisfun.com/percentage-difference.html

INTEREST

Interest (An Introduction)
Read an introduction to interest, and then click on the questions at the bottom of the page to practice.
https://www.mathsisfun.com/money/interest.html

Quiz: Simple Interest
A quiz with five questions that ask for the interest earned, final balance, interest rate, or the principal.
https://www.cliffsnotes.com/study-guides/algebra/algebra-ii/word-problems/quiz-simple-interest

Simple Interest
Another quiz where you need to find the principal, the amount of time, interest earned, or the final amount in an account earning interest. Four out of nine questions have to do with terminology and the rest are math problems.
http://www.proprofs.com/quiz-school/story.php?title=simple-interest

Simple Interest Game
Answer questions about simple interest by clicking on the correct denominations in the cash register.
http://www.math-play.com/Simple-Interest/Simple-Interest.html

Simple Interest Quiz
Test your knowledge of simple interest with this interactive self-check quiz.
https://www.proprofs.com/quiz-school/story.php?title=simple-interest

Simple Interest Practice Problems
Practice using the formula for simple interest in this interactive online activity.
http://www.transum.org/Maths/Activity/Interest/

Calculating simple interest
This page includes several video tutorials plus a short three-question quiz on simple interest.
https://www.sophia.org/concepts/calculating-simple-interest

Compound interest
A simple introduction to compound interest with many examples.
http://www.mathsisfun.com/money/compound-interest.html

GENERAL

Percents Quiz
Review basic percent calculations with this short multiple-choice quiz.
http://www.phschool.com/webcodes10/index.cfm?wcprefix=bja&wcsuffix=0605&area=view

Percent Quiz
Practice determining what percentage one number is of another in this interactive online quiz.
https://www.thatquiz.org/tq-3/?-j1c0-l7-mpnv600-p0

Math at the Mall
Practice percentages while shopping at a virtual mall. Find the percentage of discount and the sales price, calculate the interest earned at the bank, compare health memberships at the gym and figure out how much to tip your waiter at the Happy Hamburger.
http://www.mathplayground.com/mathatthemall2.html

Percent Word Problems
Practice solving word problems involving percents in this interactive online activity.
https://www.khanacademy.org/math/algebra-basics/basic-alg-foundations/alg-basics-decimals/e/percentage_word_problems_1

Percents Quiz 2
Practice answering questions about percent in this multiple-choice online quiz.
http://www.phschool.com/webcodes10/index.cfm?wcprefix=bja&wcsuffix=0606&area=view

Percentages Multiple-Choice Test
Test your understanding of percentages with this self-check multiple choice quiz.
http://www.transum.org/Software/Pentransum/Topic_Test.asp?ID_Topic=28

Percentage Quiz
Reinforce your skills with this interactive multiple-choice quiz.
https://www.mathopolis.com/questions/q.php?id=877&site=1&ref=/percentage.html&qs=877_878_879_1301_1302_880_1303_1304

Practice on Significant Figures
A multiple-choice quiz that also reminds you of the rules for significant digits.
https://web.archive.org/web/20161013062458/http://www.chemistrywithmsdana.org:80/wp-content/uploads/2012/07/SigFig.html

Review: Percent

Percent (or **per cent**) means *per hundred* or "divided by a hundred." (The word "cent" means one hundred.) So, simply put, percent means a hundredth part.

To convert percentages into fractions, simply read the "per cent" as "per 100." Thinking of hundredths, you can also easily write them as decimals.

Therefore, 8% = 8 per cent = 8 per 100 = 8/100 = 0.08.

Similarly, 167% = 167 per 100 = 167/100 = 1.67.

$$\frac{5}{100} \text{ five per cent} = 5\%$$

1. Write as percentages, fractions, and decimals.

a. 52% = ▢ = _____	b. ____% = ▢ = 0.07	c. ____% = $\frac{59}{100}$ = _____
d. 109% = ▢ = _____	e. ____% = $\frac{382}{100}$ = _____	f. 200% = ▢ = _____

A decimal number with two decimal digits is in hundredths, so it can easily be written as a percentage. For example, 0.56 = 56%. But even if we have 3 or more decimals, we can still convert into percent.

Example 1. The number 0.564 is 564 thousandths. As a percentage, 0.564 = 56.4%. Compare this to 0.56 = 56%. The decimal digit "4" that follows the digits "56" is in the thousandths place, so it becomes 4 tenths of a percent (56.4%).

$$0.091 = 9.1\% = \frac{91}{1000}$$

$$0.387 = 38.7\% = \frac{387}{1000}$$

This is how to convert percentages with even more decimal digits:

decimal	percentage		decimal	percentage
0.38429 =	38.429%		3.0281930 =	302.81930%

Think of it this way. Since 0.38 = 38%, any decimal digits that we have beyond 0.38 (the digits 429) simply become decimal digits for the percentage. In effect, we move the decimal point two places to the right.

2. Write as percentages, fractions, and decimals.

a. 28.2% = ▢ = _____	b. 6.7% = ▢ = _____	c. ____% = ▢ = 0.891
d. 0.9% = ▢ = _____	e. ____% = $\frac{1039}{10000}$ = _____	f. ____% = $\frac{3409}{1000}$ = _____
g. 45.39% = 0._____	h. 2.391% = 0._____	h. ____% = 0.942834

Writing fractions as percentages

Sometimes you can easily convert a fraction to an equivalent fraction with a denominator of 100. After that it's easy to write it as a decimal and as a percentage.

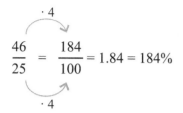

$$\frac{46}{25} = \frac{184}{100} = 1.84 = 184\%$$

For most fractions, you need to divide in order to convert them into decimals first and then into percentages.

Simply treat the fraction line as a division symbol and divide. You will get a decimal. Then write the decimal as a percentage.

Example 1.

$$\frac{8}{9} = 0.888... \approx 0.889 = 88.9\%$$

```
   0.8888
 9)8.0000
  -72
    80
   -72
    80
   -72
    80
   -72
     8
```

3. Write the fractions as percentages.

a. $\frac{8}{25} = \frac{}{100} = \underline{}\%$

b. $\frac{142}{200} = \frac{}{100} = \underline{}\%$

c. $\frac{24}{20} = \frac{}{100} = \underline{}\%$

4. Write as percentages. Use long division. Round your answers to the nearest tenth of a percent.

a. 11/8

b. 11/24

5. Use a calculator to convert the fractions into decimals. Round the decimals to four decimal digits. Then write the decimals as percentages.

a. $\frac{2}{3} \approx \underline{\ 0.6667\ } = \underline{\ 66.67\ }\%$

b. $\frac{6}{7} \approx \underline{} = \underline{}\%$

c. $\frac{17}{23} \approx \underline{} = \underline{}\%$

d. $\frac{52}{98} \approx \underline{} = \underline{}\%$

If you are asked the percentage:	**Example 3.** What percentage is 14 km of 75 km?
Asking what percent(age) is essentially the same as asking "what part" or "what fraction."	1. Write the fraction *part/total*: it is 14/75.
1. Simply write the fraction $\frac{part}{total}$.	2. Then use a calculator and write it as a decimal. $14/75 = 0.18\overline{6}$. Now write the decimal as a percentage: $0.18\overline{6} = 18.\overline{6}\%$.
2. Convert the fraction into a decimal, and then into a percentage.	Normally, we round the result and say that 14 km is about 19% of 75 km.

6. The circle graph shows the areas of the world's five oceans in square kilometers. The total area of these oceans is 335,258,000 km². To the nearest tenth of a percent, find how many percent each ocean is of the total area of the oceans.

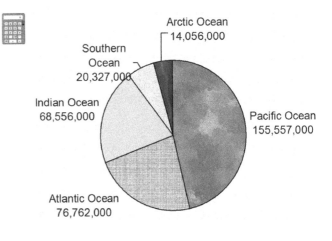

Ocean	Percentage of total area
Pacific Ocean	
Atlantic Ocean	
Indian Ocean	
Southern Ocean	
Arctic Ocean	

7. The Carters live on a rectangular piece of land that measures 40 m × 35 m. The Joneses live on a rectangular piece of land that measures 42 m × 39 m. To the nearest tenth of a percent, find what percentage the area of the Carters' land is of the area of the Joneses' land.

8. Harry has two roosters, named Captain and Chief. The weight of Captain is 7/5 of the weight of Chief.

   a. Write the second sentence above using a percentage instead of a fraction.

   b. If Chief weighs 6 lb, how much does Captain weigh?

What percentage of each figure is colored?

a.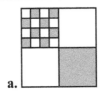

b.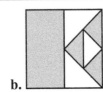

Solving Basic Percentage Problems

If the percentage is known and the total is known: (What is x% of y?)	If you are asked the percentage:
1. Write the percentage as a decimal. 2. Multiply that decimal by the total. Or use mental math tricks for finding 1%, 10%, 20%, 30%, 25%, 50%, 75%, *etc.* of a number.	Asking "what percentage" is essentially the same as asking "what part" or "what fraction." 1. Write the fraction $\frac{part}{total}$. 2. Convert that fraction into a decimal, and then into a percentage.

Example 1. A shirt that cost $34 was discounted by $4. What is the percentage it was discounted?

We write the fraction $\frac{part}{total}$ and get $\frac{\$4}{\$34}$ = 2/17 = 0.1176471 ≈ 11.8%.

Example 2. Find 59.2% of $2,600.

Write 59.2% as 0.592, and translate the word "of" into multiplication. We get 0.592 · $2,600 = $1,539.20.

Example 3. A meal that cost $14 was increased by 20%. What is the new price?

Because the price is increasing, the new price isn't 20% but 100% + 20% = 120% of the original one. So we can rewrite the previous sentence as: The new price = 1.2 · $14 = $16.80.

In this case, since the numbers are easy, we could also use mental math. Ten percent of $14 is $1.40, so the 20% increase is $2.80. The sum of the original price and the increase gives the new price: $14 + $2.80 = $16.80.

1. Change the percentages into decimals.

a. 107%	**b.** 16.67%	**c.** 4.5%

2. Calculate the new, increased prices. Write the percentages as decimals and use multiplication.

   a. Laptop: Original price $249.90, increase 6%.

   New price = _____ · $249.90 = _____

   b. Biology textbook: Original price $82.40, increase 2.5%.

   New price = _____ · $82.40 = _____

   c. Lunch buffet: Original price $18.50, increase 11.2%.

   New price = _____ · $18.50 = _____

3. Julia paid $325.08 of her $1,890 paycheck in taxes. What percentage of her paycheck did she pay in taxes?

4. Fill in Gloria's solution to the following problem.

Roller blades originally cost $45.50, but now they are discounted by 13%. What is the new price?

> Since 13% of the price is removed, _____ % of the price is left. I write that percentage as
> a decimal and multiply the original price by it: _____ · $45.50 = _____.

5. Calculate the discounted prices. Write the percentages as decimals and use multiplication.

   a. Subscription to a magazine: Original price $78, discount 38%.

   new price = _____ · $78 = _____

   b. Swimming goggles: Original price $14.95, discount 22.5%.

   new price = _____ · $14.95 = _____

6. The two rectangles are similar.

   a. In what ratio are the corresponding sides of the rectangles?

   b. In what ratio are their areas?

   c. Calculate what percentage the area of the smaller rectangle is of the area of the larger rectangle.

Sales Tax and Mental Math

A ticket to a circus costs $25. The sales tax is 6%. What is the final price you have to pay?

The sales tax is always added to the **base price** (the price without tax).
We simply calculate 6% of $25 and add that amount to $25.

Calculate in your head: 1% of $25 is $0.25. Six times that is $1.50. So the final price is $26.50.

---

7. Find the final price when the base price and sales tax rate are given. This is a mental math workout, so do not use a calculator!

a. Bicycle: $100; 7% sales tax.	b. Fridge: $400; 6% sales tax.	c. Haircut: $50; 3% sales tax.
Tax to add: $_____	Tax to add: $_____	Tax to add: $_____
Price after tax: $_____	Price after tax: $_____	Price after tax: $_____

8. Here is another mental math workout! Once again, do not use a calculator. The sales tax is 5%. For each set of items find the price with tax. *Hint: To find 5% of a number, first find 10% and take half of that.*

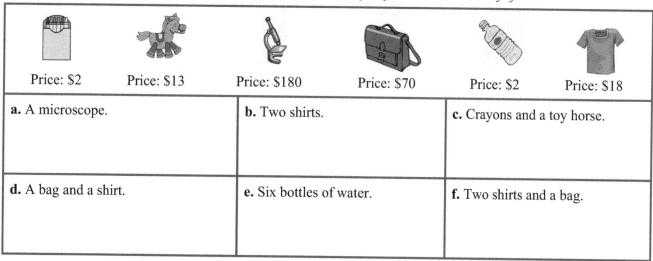

a. A microscope.	**b.** Two shirts.	**c.** Crayons and a toy horse.
d. A bag and a shirt.	**e.** Six bottles of water.	**f.** Two shirts and a bag.

9. Find the final price of a music CD with a base price of $11.50 when the sales tax is 6.7%.

10. Jeremy gets a 37.5% discount on a vacation package that normally costs $850. Find what Jeremy will pay for the vacation package.

11. Patrick bought 5,000 m² of land for a base price of $200,000. A 1.2% sales tax was added.

   a. Find the total price Patrick paid.

   b. Later, Patrick decides to sell 2,000 m² of the land to a neighbor. What should Patrick charge his neighbor in order to break even on what he paid for the part he's selling?

12. **a.** The base price of a music CD is $12.50. It is first discounted by 20% and then a 7% sales tax is added. What is the final price of the CD?

   b. The base price of a pair of jeans is $55.97. They are first discounted by 40% and then a 5% sales tax is added. What is the final price of the jeans?

13. Roger compared the unit prices of four different kinds of pasta. One kind cost $2, another $1.50, another $2.20, and another $1.70.

   a. Find the average price of the four types of pasta.

   b. If each type of pasta were discounted by 10%, then what would the average price be?

77

Percent Equations

Example 1. A handbag has been discounted by 23%, so now it costs $43.81. What was its original price?

Solution with an equation:	To solve the equation, simply divide both sides by 0.77:
Let p be the original price. A discount of 23% means that 23% of the price (or $0.23p$) is taken away from the price (p). As an expression, the discounted price is therefore $p - 0.23p$, which simplifies to $0.77p$. You can also reason that, after the discount, 77% of the price is left, so the discounted price is $0.77p$. Since the discounted price is $43.81, the equation to solve is $0.77 \cdot p = \$43.81$	$0.77p = \$43.81 \quad \vert \div 0.77$ $p \approx \$56.90$ Check: $0.77 \cdot \$56.90 \stackrel{?}{=} \$43.81$ $43.813 \approx \$43.81$ ✓
Solution with logical (proportional) reasoning: Again, we start out by reasoning that 77% of the price is left. In the chart on the right, the long lines mean "corresponds to" (not "equals").	77% —— $43.81 1% —— $43.81/77 100% —— $43.81/77 · 100 100% —— $56.90

1. Write an expression for the final price using a decimal for the percentage.

   a. Headphones: price $12, discounted by 24%. New price = _____

   b. A bag of dog food: price p, discounted by 11%. New price = _____

   c. Pizza sauce: price x, discounted by 17%. New price = _____

   d. Sunglasses: price s, price increased by 6%. New price = _____

2. A computer is discounted by 25%, and now it costs $576. Let p be its price before the discount. Select the equation that matches the statement above and solve it.

   $p + 0.25p = 576$

   $p - 0.25p = 576$

   $0.25p = 576$

3. The rent was increased by 5% and is now $215.25. What was the rent before the increase? Write an equation for this situation and solve it.

4. A tablet is discounted by 30%. Matthew bought two of them, and he paid $98. Find the price of the tablet before the discount (p).

   a. Find the equation on the right that matches the problem.

   b. Solve it.

   $2(p - 30) = 98$

   $2p - 30 = 98$

   $0.7p = 98$

   $2(p - 0.3p) = 98$

   $2(p - 0.3) = 98$

5. a. Write an expression for the final price of a property with a base price of p when a 6.5% sales tax *and* an 0.85% property tax are added to the base price.

   b. Let's say that the final price of the property in this situation is $16,639.25. How much is the base price without the taxes?

Percent proportion

Since percentages are fractions, we can easily write proportions to solve percent problems. The basic idea is to write the fraction *part/total* and set that equal to the percentage, which is written as a fraction with a denominator of 100.

That way we get this proportion: $\dfrac{part}{total} = \dfrac{percent}{100}$.

Let's solve the problem from the beginning of this lesson using a percent proportion.

Example 2. A handbag has been discounted by 23%, so now it costs $43.81. What was its original price? The discounted price of $43.18 is the "part" and the original price is the "total" (and is unknown). So we get the fraction $43.81/p. The percentage 77% is written as the fraction 77/100. We get the proportion: $$\dfrac{\$43.81}{p} = \dfrac{77}{100}$$ Its solution is on the left.	$\dfrac{\$43.81}{p} = \dfrac{77}{100}$ $77p = 100 \cdot \$43.81$ $77p = \$4{,}381$ $\dfrac{77p}{77} = \dfrac{\$4{,}381}{77}$ $p = \$56.90$
Example 3. Calculate 45% of 0.94 liters. The total is 0.94 liters, and the part is unknown. We get the percent proportion $$\dfrac{x}{0.94 \text{ L}} = \dfrac{45}{100}$$ Solve it, and verify that you get $x = 0.423$ liters.	**Example 4.** The other way to calculate 45% of 0.94 liters is to use decimal multiplication: it is $0.45 \cdot 0.94 \text{ L} = 0.423 \text{ L}$. (Simply converting percentages into decimals is often more efficient than setting up a percent proportion.)

6. A fan is discounted by 22%, and now it costs $28. Let *p* be its price before the discount.

   a. Find the proportion on the right that matches the problem.

   b. Solve it.

> $28/p = 78/100$
>
> $p/28 = 78/100$
>
> $p/78 = 28$

7. Write and solve a percent proportion (according to the data below) in the form $\frac{part}{total} = \frac{percent}{100}$.

   a. How much is 56% of 4,500 km?

   b. Thirty-eight percent of a number is 6.08. What is the number?

8. Alice bought 5 bottles of hair conditioner when the store had it at 15% off. Her total bill was $50.79. What was the price of one bottle of hair conditioner before the discount?

9. Matthew has to pay an annual property tax that is 0.8% of the accessed value (the official value for tax purposes) of his land. The tax agency told him the tax is $95.20. From that information, Matthew calculated the accessed value. What is the accessed value?

10. The price of electricity was lowered by 5%, so now it is $0.133 per kilowatt-hour. What was the price before the decrease?

11. A store owner was planning for a big 30% off sale. However, she was rather unethical about it, and the night before the sale, she increased the prices on some of the sale items. For example, she increased the price by 30% for a roll of ribbon that did cost $5. What will the sale price of this roll of ribbon be?

12. The area of a triangular piece of land is 6 square kilometers. If the dimensions, including the base and the altitude, of the triangle were increased by 10%, by how many percent would its *area* increase?
    Hint: Make up a triangle with the given area. In other words, make up a base and an altitude so that the area is 6 km².

Puzzle Corner

A family's water bill for the whole year was $584. From August through December, the bill was 10% higher than from January through July because of a 10% price increase. What was the monthly water bill before the increase?

Circle Graphs

A **circle graph** shows visually how a total is divided into parts (percentages). Each of the parts (pie slices) is a **sector**, and each sector has a **central angle**.

To make a circle graph, we need to calculate the measure of the central angle of each sector. For example, if a circle graph is supposed to show the percentages 25%, 13%, and 62%, simply calculate those percentages of 360° (the full circle):

25% of the total corresponds to 0.25 · 360° = 90°.
13% of the total corresponds to 0.13 · 360° = 46.8°.
62% of the total corresponds to 0.62 · 360° = 223.2°.

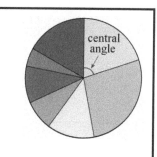

1. Sketch a circle graph that shows...

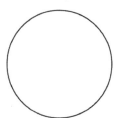
a. 50%, 25%, and 25%

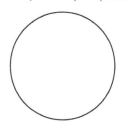
b. 33.3%, 33.3%, 1/6, and 1/6

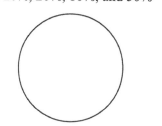
c. 20%, 20%, 10%, and 50%

2. The table shows different kinds of specialty breads that a grocery store ordered. Fill in the table. Make a circle graph. (*Note:* You will need a protractor to draw the angles.)

Type	Quantity	Percentage	Central Angle
white bread	50		
bran bread	25		
rye bread	30		
corn bread	40		
4-grain bread	55		
TOTALS	**200**	**100%**	**360°**

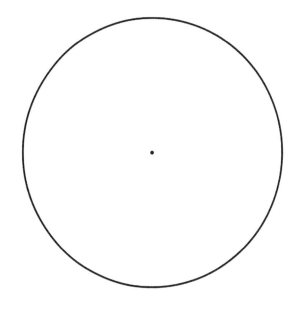

3. **a.** Make a bar graph of the quantities of each type of bread from the table above. →

   b. Does the bar graph show percentages?

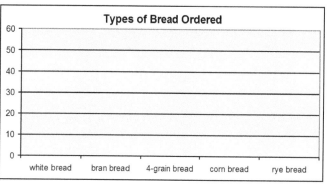

83

4. Think of fractions. Estimate how many percent the sectors of the circle graphs represent.

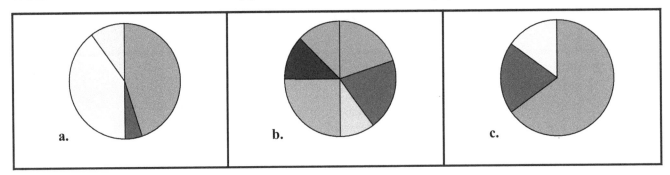

a.

b.

c.

5. The table lists by flavor how many units of protein powder a company sold. Draw a circle graph showing the percentages. You will need a protractor and a calculator.

Flavor	Amount sold	Percentage of total	Central Angle
chocolate	67		
vanilla	34		
strawberry	16		
blueberry	26		
TOTALS		100%	360°

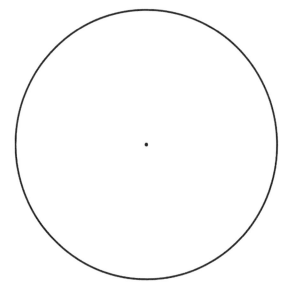

6. Mark polled some seventh graders about their favorite hobbies. Below are his results. Draw a circle graph to show the percentages. Round the angles to whole degrees. You will need a protractor and a calculator.

Favorite hobby	Percentage	Central Angle
Reading	12.3%	
Watching TV	24.5%	
Computer games	21%	
Sports	22.3%	
Pets	7.1%	
Collecting	8.1%	
no hobby	4.7%	
TOTALS	100%	360°

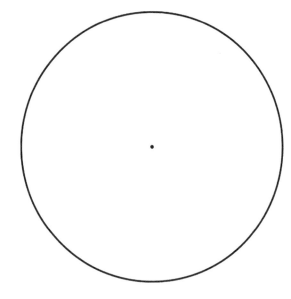

84

Percentage of Change

Percent(age of) change is a way to describe how much a price or some other quantity is increasing or decreasing (changing). Let's look at how to calculate the percentage a quantity is changing.

Example 1. A phone used to cost $50. Now it has been discounted to $45. What percentage was the discount?

Since this problem is asking for the *percentage*, we will use our basic formula $\frac{part}{total} = percentage$.

Because the change is relative to the *original* price, that original price becomes the "total" in our equation. The "part" is the actual amount by which the quantity changes, in this case $5. So we get

$$percentage = \frac{part}{total} = \frac{\$5}{\$50} = 1/10 = 10\%$$

Essentially, we wrote **what fraction the $5 discount is of the original $50 price** and converted that fraction into a percentage.

In summary: To calculate the percent change, use the same basic formula that defines a percentage: *part/total*. Since the change is relative to the original price, the original price is the "total," and the change in price is the "part."

$$\text{percentage of change} = \frac{part}{total} = \frac{difference}{original}$$

1. Write an equation and calculate the percentage of change.

a. A toy construction set costs $12. It is discounted and costs only $8 now. What percentage is the discount?

$$\frac{difference}{original} =$$

b. A sewing kit costs $20. It is discounted and costs only $16 now. What percentage is the discount?

c. A bouquet of flowers used to cost $15, but now it costs $20. What is the percentage of increase?

d. The price of a stove was $160. The price has increased, and now it costs $200. What is the percentage of increase?

Compare these two problems:	
Gasoline cost $3/gallon last week. Now it has gone up by 5%. What is the new price?	Gasoline cost $3/gallon last week. Now it costs $3.15. What was the percentage of increase?
1. Calculate 5% of $3. Since 10% of $3 is $0.30, we know that 5% is half of that, or $0.15. 2. Add $3 + $0.15 = $3.15/gallon. That is the new price.	1. Find how much was added to $3 to get $3.15 (the difference). That is $0.15. 2. Find what percentage $0.15 is of the original price, $3. It is 15/300 = 5/100 = 5%. So the percentage of increase was 5%.
To find the percentage of increase (the right box above), we work "backwards" compared to when we find the new price when the percentage of increase is known (the left box above).	

2. Solve and compare the two problems.

a. A shirt used to cost $24 but it was discounted by 25%. What is the new price?	**b.** A shirt used to cost $24. Now it is discounted to $18. What percentage was it discounted?

3. Solve and compare the two problems.

a. At 5 months of age, a baby weighed 5 kg. At 6 months, the baby weighs 6 kg. What was the percentage of increase?	**b.** At 5 months, a baby weighed 6 kg. Over the next month, his weight increased by 20%. What is his weight at 6 months of age?

4. From June to July, the rent increased from $325 to $342. Then it increased again in August, to $349. Which increase was a greater percentage?

5. **a.** The price of a biology textbook was $60. Then it was lowered to $54. Calculate the percentage change in price.

   b. The price was increased back to $60. Calculate the percentage of increase. Hopefully this is not a surprise to you, but the percentage is *not* the same as in part (a)!

6. A jacket cost $50. First its price was increased by 20%. Then it was discounted by 20%.

   a. Calculate the final price. It will *not* be $50!

   b. Since the original price was $50, use your answer from part (a) to calculate the true overall percentage change in price.

7. Jake's work hours were cut from 40 to 37.5 a week. Anita's work hours were cut from 145 to 135 a month. Whose work hours were cut by a greater percentage?

8. The price of a vacuum cleaner that cost $100 is increased by 20%. Then it is increased by another 10%.

   a. Find the price of the vacuum cleaner now.

   b. Find the percentage of increase if the price had been increased from $100 to the final price in one single increase. Note: the answer will *not* be 30%!

Percentage of Change: Applications

Area
Example 1. A children's playground measures 30 ft × 40 ft. It is enlarged so that each side is 10 ft longer. What is the percentage of increase in the area?   The question doesn't ask for the percent increase of each *side*, but of the *area*.   The original area is 30 ft × 40 ft = 1,200 sq. ft. The new area is 40 ft × 50 ft = 2,000 sq. ft.   Now we can find the percentage of increase. The fraction *difference/original* is (800 sq. ft. / 1,200 sq. ft) = 8/12. In lowest terms 8/12 becomes 2/3, which, as a percentage, is 66.7%.

Give your answers to the nearest tenth of a percent.

1. Find the percentage of increase in area when a 10 m × 10 m garden is enlarged to be 15 m × 15 m.

2. A newsletter has been printed on 21 cm × 29.7 cm paper. To save costs, it will be printed on 17.6 cm × 25 cm paper instead. By how many percent will the printable area decrease?

3. The sides of a square are increased by a scale factor of 1.15.

    a. By what percentage does the length of each side increase?

    b. What is the percentage of increase in area?
      Hint: Make up a square using an easy number for the length of side.

    c. (Challenge) Would your answers to (a) and (b) change if the shape were a rectangle? A triangle?

4. The graph below compares the production of water in California for usage in June, July, and August of 2013 to the production in the corresponding months of 2014. Because of imposed restrictions on water usage, less water was used in 2014 than in 2013.

   Notice the scale is in millions of gallons. For example, in June, 2013, California produced 215,363 *million* gallons of water, not 215,363 gallons.

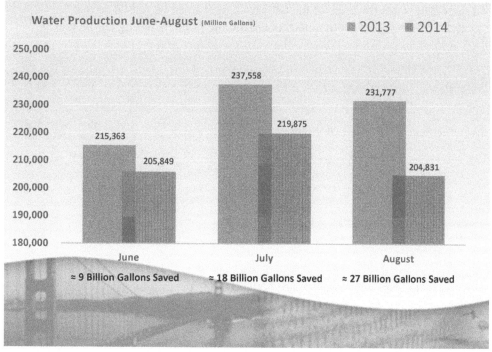

Source: California Water boards

   a. Why do you think the water production (and usage) is higher in July than in June?

   b. Calculate the percentage of decrease between 2013 and 2014 for each of the three months.

   c. In which month was the greatest percentage decrease in water production?

   How can you see that from the graph?

5. The price of a jar of honey went from $5.50 to $6.00. Then it increased further to $6.50. If the price were to increase by another $0.50 (from $6.50 to $7.00), would the percentage increase be more than, less than, or the same as when the price was increased the first time?

6. The sales tax is 7%. The price of a solar battery charger *with tax* is $69.99.

   a. What is the price without tax?

   b. Let's say a merchant wants the price of the battery charger to be increased so that the price including tax is $79.99. What percentage did the price including tax increase?

   c. What is the percentage of increase of the price before tax?

7. A designer plans to use windows of the size 85 cm × 85 cm.
   He changes his mind and uses windows that are both 10 cm wider and longer instead.

   a. By how many percent does that change the area of one window?

   b. Originally, the designer was going to use 20 of the smaller windows in the house. About how many of the bigger windows cover the same area as 20 of the smaller ones?

8. a. Three items, with prices of $50, $60, and $70, have their prices *increased* by $10. For which item is the *percentage* increase in price the greatest?

   b. Three items, with prices of $50, $60, and $70, have their prices *discounted* by 12%. Which item's price decreases the most (in dollars)?

9. The population of the state of Kentucky was 3,038,000 in 1960 and 3,219,000 in 1970. Calculate the percentage of increase in the population of Kentucky during that decade.

10. The table shows the population of Kentucky every 10 years. The line graph shows the same information. Your task is to calculate the percentage of increase in each decade. You already did that for 1960-1970 in the previous exercise, so write that answer in the row for 1970.

Year	Population	% increase in the decade
1960	3,038,000	N/A
1970	3,219,000	
1980	3,661,000	
1990	3,685,000	
2000	4,042,000	
2010	4,340,000	

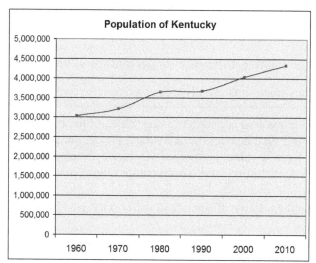

11. The population of Kentucky grew more steeply in one of these decades than any of the others.

   a. Which decade? From _____ to _____.

   b. How can you tell which decade it is from the graph?

12. (Optional.) Make a line graph that shows the change in the population where you live (your town, city, parish, county, state, province, country, *etc.*). Also calculate the percentage of increase or decrease for each time period. You can often find population statistics in Wikipedia, for example. You can also search the Internet (with adult supervision) for "your area population statistics," where your area is the place that you chose.

Comparing Values Using Percentages

What percentage more/less/bigger/smaller/taller/shorter ...?

Example 1. A car weighs 2,000 kg. Another car weighs 2,500 kg. The second car is heavier, but how much heavier than the first one is it?

To answer that question, we cannot just look at the difference of 500 kg and say that 500 kg is "a lot" or "a little." Instead, we need to take into account the sizes of the two items being compared and consider their *relative difference*.

The **relative difference** (or **percentage of difference**) of two values is the fraction $\frac{\text{actual difference}}{\text{reference value}}$.

This fraction is usually expressed as a percentage.

The problem is in determining which of the two values is the reference value. In this case, we want to find out how much heavier the second car is than the first car, which means we use the weight of the first car as the reference value. In another words, we are comparing the second car to the first car. It is as if the lighter car were there first, and we are comparing a "newcomer" car to this first car.

So the relative difference of the two weights is $\frac{500 \text{ kg}}{2{,}000 \text{ kg}} = \frac{5}{20} = \frac{1}{4} = 25\%$.

This means the second car is 25% heavier than the first one. This gives us a precise value.

Choosing the second car's weight as the reference value gives: $\frac{500 \text{ kg}}{2{,}500 \text{ kg}} = \frac{5}{25} = \frac{1}{5} = 20\%$.

This means the first car is 20% lighter than the second car.

Example 2. Southcreek College has 2,600 students and West River College has 2,400 students. How many percent more students does Southcreek College have than West River College?

The difference in student count is 200. But which number is our reference? Since West River College is mentioned after the word "than," we are comparing Southcreek College to West River College. So the student count of West River College is our reference number.

The fraction is $\frac{\text{difference in student count}}{\text{reference student count}} = \frac{200}{2{,}400} = \frac{2}{24} = \frac{1}{12} = 0.08333... \approx 8.3\%$.

So Southcreek College has approximately 8.3% more students than West River college.

---

1. Compare the taller object to the shorter one and calculate their relative difference.

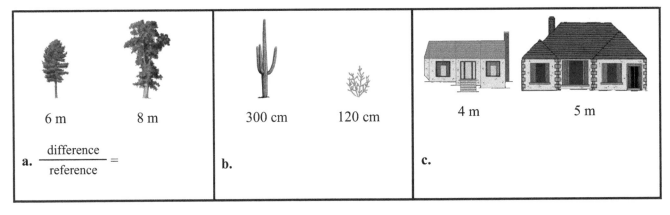

a. $\frac{\text{difference}}{\text{reference}} =$

b.

c.

You may use the calculator in all problems in this lesson from this point on.

2. Erica is 140 cm tall, and Heather is 160 cm tall. Fill in the blanks.

   Heather is $\dfrac{\quad}{\quad}$ = _____% taller than Erica. Erica is $\dfrac{\quad}{\quad}$ = _____% shorter than Heather.

3. **a.** Refer to the chart. Calculate the percentage differences to the tenth of a percent.

   The population of Tokyo is _____ % larger than the population of Delhi.

   The population of Moscow is _____ % smaller than the population of Mexico City.

   b. Compare how much larger the population of Seoul is than the population of New York, and then again how much larger the population of New York is than the population of Moscow.

   Which is a larger percentage difference?

Metropolitan area	Country	Population (millions)
Tokyo	Japan	37.80
Seoul	South Korea	25.62
Shanghai	China	24.75
Karachi	Pakistan	23.50
Delhi	India	21.75
Mexico City	Mexico	21.60
Sao Paulo	Brazil	21.20
Jakarta	Indonesia	20.00
New York City	United States	19.95
Mumbai	India	20.75
Moscow	Russia	15.51

When there is no clear way to choose a reference value when calculating the relative (percentage) difference, you can use the average of the two values as a reference value.

Example 3. The monthly subscription fees to two math practice websites are $9.99 and $12.95. What is their relative difference?

The average of the two prices is ($9.99 + $12.95)/2 = $11.47.

The relative difference is therefore $\dfrac{\$12.95 - \$9.99}{\$11.47} = \dfrac{2.96}{11.47} \approx 25.8\%$.

---

4. The area of one park is 14,000 square feet, and the area of another park is 10,000 square feet. Use their average area to calculate the relative percentage difference between their areas.

5. KeepCool company charges $28 per hour for labor, and CityFreez charges $32 per hour.

   a. Use the average rate for labor to calculate the percentage difference in the rates.

   b. Now compare the costs for *two* hours of labor. What is the percentage difference?

Example 4. One cat weighs 1.2 kg and another weighs 1.5 kg.

	Compare carefully these five types of questions about the same situation!

Question 1. What percentage is the smaller cat's weight <u>of</u> the bigger cat's weight?

Solution: Write what fraction the smaller cat's weight is relative to the bigger cat's weight:

It is $\dfrac{\text{smaller cat's weight}}{\text{bigger cat's weight}} = \dfrac{1.2 \text{ kg}}{1.5 \text{ kg}} = \dfrac{12}{15} = \dfrac{4}{5} = 80\%.$

Question 2. What percentage is the bigger cat's weight <u>of</u> the smaller cat's weight?

Solution: Write what fraction the bigger cat's weight is relative to the smaller cat's weight:

It is $\dfrac{\text{bigger cat's weight}}{\text{smaller cat's weight}} = \dfrac{1.5 \text{ kg}}{1.2 \text{ kg}} = \dfrac{15}{12} = \dfrac{5}{4} = 1 \tfrac{1}{4} = 125\%.$

Question 3. How much more (in percent) does the bigger cat weigh <u>than</u> the smaller cat?

Solution: This is a percentage difference relative to the smaller cat. Write the fraction (difference / reference):

It is $\dfrac{\text{difference in weight}}{\text{smaller cat's weight}} = \dfrac{0.3 \text{ kg}}{1.2 \text{ kg}} = \dfrac{3}{12} = \dfrac{1}{4} = 25\%.$

Question 4. Compare the weight of the smaller cat <u>to</u> the weight of the bigger cat.

Solution: This percentage difference is relative to the weight of the bigger cat:

It is $\dfrac{\text{difference in weight}}{\text{bigger cat's weight}} = \dfrac{0.3 \text{ kg}}{1.5 \text{ kg}} = \dfrac{3}{15} = \dfrac{1}{5} = 20\%.$

Question 5. What is the relative difference between the two cats' weights?

Solution: No reference is specified, so let's use the average weight: ½(1.5 kg + 1.2 kg) = 1.35 kg.

It is $\dfrac{\text{difference in weight}}{\text{average weight}} = \dfrac{0.3 \text{ kg}}{1.35 \text{ kg}} = \dfrac{30}{135} = \dfrac{2}{9} = 0.222... \approx 22.2\%.$

There are five different questions with five different solutions. Note the underlined key words!

You could be asked, **"What percentage is (this) <u>of</u> (that)?"**

OR: **"What percentage <u>more/less/smaller/bigger</u> is (this) <u>than</u> (that)?"**

Moreover, the order of comparison matters: The keywords "of," "than," and "to" mark the cat that we are comparing to (the reference cat).

6. One bean plant is 12 cm tall, and another is 16 cm tall.

a. What percentage is the shorter plant's height of the taller plant's height?	**b.** How much taller (in percent) is the taller plant than the shorter plant?
c. Compare the height of the shorter plant to the height of the taller one.	**d.** Find the relative difference in the heights using their average height.

7. Only one of the students gets the answer correct in each case. Who is it?

   a. Jack made a tower of blocks 150 cm tall. Baby made a block tower that was 30 cm tall. What percentage is the height of Baby's tower of the height of Jack's tower?

Elijah:	Angela:	Mary:
I subtract $150 - 30 = 120\%$.	I write the fraction $\dfrac{120 \text{ cm}}{150 \text{ cm}} = \dfrac{12}{15} = \dfrac{4}{5} = 80\%$	I write the fraction $\dfrac{30 \text{ cm}}{150 \text{ cm}} = \dfrac{3}{15} = \dfrac{1}{5} = 20\%$

   b. The school orchestra has 26 boys and 14 girls. How many percent bigger is the number of boys than the number of girls?

Elijah:	Angela:	Mary:
I subtract $26 - 14 = 12$ and write the fraction $\dfrac{12}{14} = \dfrac{6}{7} \approx 86\%$	I subtract $26 - 14 = 12\%$.	I write the fraction $\dfrac{14}{26} = \dfrac{7}{13} \approx 54\%$

8. Percentage comparisons can be misleading without the actual data. To see that, consider the number of violent crimes committed in two imaginary counties in 2013 and 2014.

   a. Calculate the percentage of increase in violent crime for the two counties in the table.

   b. Let's say you were thinking of moving into one of these two counties, and you were told *only* the percentages of change for violent crime. Which county might you move into?

Number of violent crimes in 2013-2014

	County A	County Z
2013	2	454
2014	4	512
Percentage of increase		

Yet, taking the *actual data* into account, your perception about these two areas would change a lot!

The table lists the number of domestic burglaries in three counties in Wales for two different time periods.

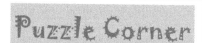

Number of domestic burglaries

	Ceredigion	Conwy	Gwynedd
April 2011 to March 2012	93	256	200
April 2012 to March 2013	56	229	138
Approximate percent of change			

Determine *without* a calculator or paper and pencil (just figure in your head) which county experienced the greatest percentage of decrease in domestic burglaries.

Simple Interest

When you deposit money into a savings account, normally the bank actually pays you for the use of your money. The amount you deposit is called the **principal**. The amount the bank pays is called the **interest**.

Similarly, if you borrow money, it is not free. You not only pay back the principal (the actual amount you borrowed) but you also pay interest on the amount of money you borrowed.

Interest is normally defined as a certain percentage, called the **interest rate**, of the principal. For example, the bank might charge you an interest rate of 7.9% on a loan. Often, people simply say "7.9% interest" instead of "a 7.9% interest rate" or "an interest rate of 7.9%."

In this lesson, we study only so-called **simple interest**, which means that the interest is added to the principal only at the end of the time period during which the money is invested or borrowed. In real life, banks usually calculate **compound interest**, which means that the interest is added to the principal at certain regular intervals (such as every month or even every day) during the period of the loan or investment.

Example 1. How much interest does a principal of $2,000 earn in a year if the yearly interest rate is 5%?

The interest is simply 5% of $2,000, which is $100.

Example 2. You get a $3,000 loan with an annual interest rate of 8.5%.
You pay the loan back after 3 years. How much do you have to pay back?

The interest for one year is simply 8.5% of $3,000. For three years, it is three times that much.
Of course you also have to pay back the of principal $3,000. So the total amount you pay back is:

$3,000 + 3 · 0.085 · $3,000 = $3,000 + $765 = $3,765.

We can see from the above examples that to calculate the interest, we simply take the interest rate times the principal times the time period. The formula for calculating simple interest (I) is usually given as:

$$I = Prt$$

where P is the principal, r is the interest rate, and t is the time.

You may use a calculator in all problems in this lesson.

1. Calculate the interest and the total amount to be paid back on these investments.

   a. Principal $5,000; interest rate 3%; time 1 year

   Interest: _____ Total to withdraw: _____

   b. Principal $3,500; interest rate 4.3%; time 4 years

   Interest: _____ Total to withdraw: _____

   c. Principal $20,000; interest rate 7.6%; time 10 years

   Interest: _____ Total to withdraw: _____

2. Sandy plans to invest $3,000 for three years. She could either put it into a savings account with an annual interest rate of 3.4% or get a Certificate of Deposit (CD) for 5 years with an annual interest rate of 3.92%. However, if she withdraws the money from the CD before 5 years is up, the bank charges her a penalty of 6 months interest. Which allows Sandy to earn more money on her investment in 3 years?

Savings Account

Interest rate: 3.4%

Certificate of Deposit

Time period: 5 years

Interest rate: 3.92%

Example 3. Andy borrows $2,000 with an annual interest rate of 12.45%.
He is able to pay it back after 7 months. How much will Andy pay to the lender?

Notice that the interest rate is *annual* (yearly) but the time period is in *months*. Therefore:

(1) We need to either use a monthly interest rate. To get that, simply divide the annual interest rate by 12.

(2) Or we need to convert the time of 7 months into years. Of course, 7 months is simply 7/12 years.

Let's use the first option. The monthly interest rate is 12.45% ÷ 12 = 1.0375%. Then, the interest is 0.010375 · 7 · $2,000 = $145.25. So Andy has to pay back $2,145.25.

---

3. Elizabeth bought a tablet for $450 on credit with a 12.9% annual interest rate.

   a. How much interest (in dollars) will she pay in a month?

   b. In a day?

4. A credit card has a monthly interest rate of 1.09%, which doesn't sound like much. How much interest will you pay if you purchase a couch for $690 with the credit card and pay it back after two years?

5. Jerry took out a loan for $850 for 10 months with an annual interest rate of 10.8%. How much less interest would he have paid if instead he had taken out a loan for 7 months with an annual interest rate of 9.5%?

6. John uses his credit card to finance a car for $26,000. The annual interest rate on his card is low, 2.75%, but only for the first 12 months. After that, if Jon hasn't paid the full amount back, the annual interest rate jumps to 9.95%. Calculate how much John ends up paying back if he cannot pay the total during the first 12 months, but pays the entire amount after 2.5 years.

Example 4. Eric borrowed $1,500 for 8 months and paid back $1,578. What was the interest rate?

The actual interest was $1,578 − $1,500 = $78. The loan was for 8 months, so the monthly interest was $78 ÷ 8 = $9.75.

This amount is $9.75/$1,500 = 0.0065 = 0.65% of the principal. So the monthly interest rate was 0.65%. The annual interest rate is 12 times that, or 7.8%.

Another solution. We could also write an equation, using the formula $I = Prt$.

This formula is for the actual amount of interest (I), not the total Eric paid back. The actual interest (I) was $78. We also know that the time was 8 months and that the principal was $1,500. Substituting those values into the formula and keeping the units, we get

$$\$78 = \$1,500 \cdot r \cdot 8 \text{ months}$$

It helps to write the equation so that the variable is on the left (inverting the sides):

$$\$1,500 \cdot r \cdot 8 \text{ months} = \$78$$

To solve this equation for r, we divide both sides by $1,500 and then also by 8 months. See the solution on the right.

$$\$1,500 \cdot r \cdot 8 \text{ months} = \$78$$

$$\frac{\$1,500 \cdot r \cdot 8 \text{ months}}{\$1500} = \frac{\$78}{\$1500}$$

$$r \cdot 8 \text{ months} = \frac{78}{1500}$$

$$\frac{r \cdot 8 \text{ months}}{8 \text{ months}} = \frac{\frac{78}{1500}}{8 \text{ months}}$$

$$r = \frac{0.0065}{\text{month}}$$

$$r = 0.65\% \text{ per month}$$

7. **a.** You borrowed $1,000 for one year. At the end of the year, you had to pay back $1,045. What was the interest rate?

b. You borrowed $12,000 for five years. At the end of the five years you paid the bank back $15,600. What was the interest rate?

8. Alice opened a savings account that paid an interest rate of 6%. After ten years her account contained $12,000. How much was the original principal?

9. How long would you have to invest $2,000 in order to earn $500 in interest, if the annual interest rate is 11.5%?

10. What rate of interest do you need in order to earn $350 in interest in 2 years with a principal of $1,800?

11. A family purchased a vacation package for $4,055 with a credit card that charged 11.95% annual interest. When they paid off their credit card, they ended up paying the credit card company $4,741.48. How long did it take for the family to pay off the debt for their vacation?

Puzzle Corner

Let's look at just one example of how **compound interest** is calculated! Jayden buys a motorcycle for $5,000 with a credit card that has a 6% annual interest rate. Compounding the interest means that the interest is added to the principal at certain intervals, in this case each month.

Study the table below. Note that a 6% annual interest rate means $0.06/12 = 0.005$ or 0.5% monthly interest rate.

a. Fill in the table, calculating the interest for each month and adding it to the principal. The following month, use the new principal to calculate the interest.

Month	Monthly interest	Principal
0		$5,000
1	$5,000 · 0.005 = $25	$5,025
2	$5,025 · 0.005 = $25.125	$5,050.125
3	$5,050.125 · 0.005 ≈ $25.251	$5,075.376
4	_____ · 0.005 ≈	
5	_____ · 0.005 ≈	
6		
7		
8		

b. There is a formula for compound interest that allows these calculations to be done quicker. For this situation, the total amount owed after month n is

$$\$5{,}000 \cdot 1.005^n$$

For example, after 12 months, Jayden owes $\$5{,}000 \cdot 1.005^{12} = \$5{,}308.39$. Use the formula to calculate how much Jayden pays if he pays back the loan at the end of 2 years.

Chapter 7 Mixed Review

1. Evaluate the expressions. Give your answer as a fraction or mixed number.

   a. $\dfrac{3x}{x+7}$, when $x = -4$

   b. $\dfrac{1-x}{1+x}$, when $x = 7$

2. Eric walks at a constant speed of 2 m/s for eight seconds. Then he runs for the next ten seconds at a constant speed of 5 m/s.

   a. Plot a graph for the distance Eric runs.

   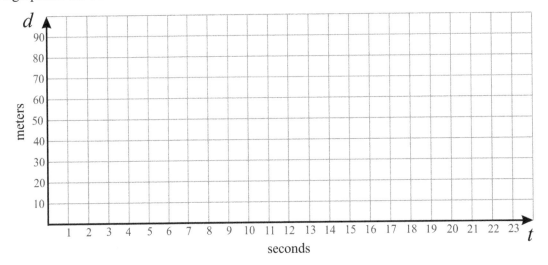

   b. What is the total distance Eric advances?

3. Solve.

   a. $\dfrac{6x}{7} = -1.2$

   b. $6x - 7 = -1.2$

4. Are the expressions equal, no matter what values x and y have? If so, you don't need to do anything else. If not, provide a counterexample.

a. $\dfrac{x-y}{3}$ $\dfrac{x}{3} - \dfrac{y}{3}$	b. $x - 2y$ $y - 2x$

5. Evaluate the expression $|a - b|$ for the given values of a and b. Check that the answer you get is the same as if you had used a number line to figure out the distance between the two numbers.

a. a is -5 and b is 6	b. a is -2 and b is -11

6. **a.** Write an expression for the distance between x and 7.

   b. Evaluate your expression for $x = -3$.

7. Write using symbols, and simplify if possible.

   a. the opposite of -2

   b. the absolute value of -80

   c. the opposite of the sum $6 + 7$

   d. the absolute value of the sum $-4 + 5$

8. Solve the proportions by using cross-multiplication.

a. $\dfrac{14 \text{ mi}}{0.59 \text{ gal}} = \dfrac{100 \text{ mi}}{V}$	b. $\dfrac{P}{2000 \text{ lb}} = \dfrac{\$4.05}{3 \text{ lb}}$

9. If gasoline costs $3.14 per gallon and if your car gets 21 miles per gallon, find the cost of driving the car for 15 miles.

10. The two figures are similar. Find the length of the unknown side.

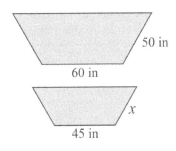

11. Solve using both decimal multiplication and fraction multiplication.

$0.6 \cdot 0.7$	**Decimal multiplication:**	**Fraction multiplication:**

12. Solve using both decimal division and fraction division.

$0.24 \div 0.5$	**Decimal division:**	**Fraction division:**

Chapter 7 Review

1. Find the percentage of the given quantity.

   a. 9.2% of $150   **b.** 45.8% of 16 m   **c.** 0.6% of 700 mi

2. All these items are on sale. Calculate the new, discounted prices.

   a.
   Price: $9
   20% off
   New price: $_____

   b.
   Price: $6
   25% off
   New price: $_____

   c.
   Price: $90
   30% off
   New price: $_____

3. A flashlight is discounted by 18%, and now it costs $23.37. Let p be its price before the discount. Find the equation that matches the statement above and solve it.

   $p - 0.18 = \$23.37$

   $p - 18 = \$23.37$

   $0.82p = \$23.37$

   $0.18p = \$23.37$

4. Two brothers, Andy and Jack, shared the price of a new computer so that Andy paid 2/5 (or 40%) and Jack paid 3/5 (or 60%) of the price. The computer cost $459, and there was a sales tax of 7% that was added onto the price. Calculate Andy's and Jack's shares.

5. A portable reading device costs $180. Now it is discounted and costs $153. What was the percentage of discount?

6. The two right triangles on the right are similar.

   a. Calculate what percentage the area of the smaller triangle is of the area of the larger triangle.

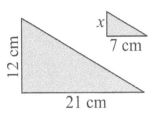

   b. In what ratio are the corresponding sides of the triangles?

   c. In what ratio are their areas?

7. A wall painting was planned to be 5 m by 3 m in size. If both of its sides are increased by 20%, by what percentage will the area of the painting increase?

8. In a race, Old Paint finished in 120 seconds, and The Old Gray Mare finished in 200 seconds.

   a. How many percent quicker was Old Paint than The Old Gray Mare?

   b. How many percent slower was The Old Gray Mare than Old Paint?

   c. What is the relative difference in their times?

9. Noah takes a $4,000 loan at a 7.8% annual interest rate to purchase a car. At the end of the first year, he pays back $2,000 of the principal of the loan. At the end of the second year, he pays back the rest of the principal and all of the interest. How much total interest does he have to pay? Assume simple interest, which is calculated and paid only at the end of the period of the loan.

Chapter 8: Geometry
Introduction

The main topics we study in this chapter are:

- various angle relationships
- drawing geometric figures, including basic geometric constructions
- pi and the area of a circle
- slicing 3-dimensional solids
- surface area and volume.

In the first lesson of the chapter, we examine various angle relationships: angles that are formed when several rays originate from the same starting point, vertical angles (formed when two lines intersect), and corresponding angles (formed when a line intersects two parallel lines). Then, the lesson *Angles in a Triangle* presents and proves the well-known result that the angles in a triangle sum to 180 degrees. With this knowledge, students are now able to solve various problems that involve unknown angles.

Next, students practice drawing geometric figures. Basic geometric constructions are done just like in ancient times: with only a compass and straightedge (a ruler without measurement units). These constructions help students to think about the main defining features of a figure. Personally I have always enjoyed geometric constructions because they are like little puzzles to solve.

Students also draw figures using a normal ruler and compass in the lesson *Drawing Problems*. They especially determine whether the given information defines a unique figure (triangle or a parallelogram).

Then we turn our attention to pi. Students first learn the definition of pi as a ratio of a circle's circumference to its diameter in the lesson *Circumference of a Circle*. Then they learn and practice how to calculate the area of a circle in a wide variety of word problems and applications. We also briefly study the proof for the formula for the area of a circle. I feel it is important that students encounter justifications for mathematical formulas and procedures and even read some proofs before high school. We don't want students to think that mathematics is only a bag of magic tricks or formulas to memorize that seemingly came out of nowhere. Proofs and logical thinking are foundations to mathematics and school mathematics should not be left without them.

After this, we slice three-dimensional solids with a plane, and learn that the result is always a two-dimensional shape. Students see that in a concrete way by slicing cubes and pyramids made of modeling clay. Some Internet links (provided in the lesson) will also help students to visualize what happens when a solid is cut with a plane.

In this chapter, students also solve a variety of problems concerning surface area and volume and practice converting between various units of area and volume. While these topics tend to involve lots of calculations and less possibilities for hands-on activities, they are very important in real life.

Consider mixing the lessons from this chapter with lessons from some other chapter, such as the chapter on probability. For example, the student could study topics from this chapter and from the probability chapter on alternate days, or study a little from each chapter each day. Such, somewhat spiral, usage of the curriculum can help prevent boredom, and also to help students retain the concepts better.

Recall also that I don't recommend automatically assigning all the problems. Use your judgment.

You can find matching videos for topics in this chapter at http://www.mathmammoth.com/videos/ (grade 7).

The Lessons in Chapter 8

	page	span
Angle Relationships	113	*5 pages*
Angles in a Triangle	118	*5 pages*

Basic Geometric Constructions	123	*6 pages*
More Constructions	129	*5 pages*
Drawing Problems	134	*7 pages*
Circumference of a Circle	141	*3 pages*
Area of a Circle	144	*3 pages*
Proving the Formula for the Area of a Circle	147	*2 pages*
Area and Perimeter Problems	149	*5 pages*
Surface Area	154	*4 pages*
Conversions Between Customary Units of Area	158	*3 pages*
Conversions Between Metric Units of Area	161	*3 pages*
Slicing Three-Dimensional Shapes	164	*7 pages*
Volume of Prisms and Cylinders	171	*4 pages*
Chapter 8 Mixed Review	175	*3 pages*
Chapter 8 Review	178	*8 pages*

Helpful Resources on the Internet

ANGLE RELATIONSHIPS

Math Warehouse - Angles
Find interactive demonstrations, examples and practice problems concerning various types of angles.
http://www.mathwarehouse.com/geometry/angle/complementary-angles.php

http://www.mathwarehouse.com/geometry/angle/supplementary-angles.php

http://www.mathwarehouse.com/geometry/angle/vertical-angles.php

http://www.mathwarehouse.com/geometry/triangles/

Angles at a Point
Drag the points of the angles to see how the angle measurements of the model change.
http://www.transum.org/software/SW/Angle_Theorems/ShowOne.asp?T=1

Angles Around a Point
A short lesson showing that angles around a point add up to 360 degrees, followed by self-check questions.
http://www.mathsisfun.com/angle360.html

Angle Points Exercises
Apply the properties of angles at a point, angles at a point on a straight line, and vertically opposite angles in this interactive online exercise.
http://www.transum.org/software/SW/Starter_of_the_day/Students/AnglePoints.asp?Level=4

Complementary & Supplementary Angles from Maths Is Fun
Each page includes a clear explanation, an interactive exploration, and self-check interactive questions.
http://www.mathsisfun.com/geometry/complementary-angles.html

http://www.mathsisfun.com/geometry/supplementary-angles.html

Complementary & Supplementary Angles
Find missing angles in this interactive exercise.
https://www.khanacademy.org/math/geometry/hs-geo-foundations/hs-geo-angles/e/complementary_and_supplementary_angles

Working with Angles
Online lessons with interactive self-check questions from Absorb Mathematics course. The lessons cover measuring angles, the types of angles, angles on a straight line, and other angle topics.
http://www.absorblearning.com/mathematics/demo/units/KCA003.html

Identifying Supplementary, Complementary, and Vertical Angles
Practice telling whether two angles are supplementary, complementary, or vertical.
https://www.khanacademy.org/math/basic-geo/basic-geo-angle/vert-comp-supp-angles/e/identifying-supplementary-complementary-vertical

Finding Missing Angles
Use your knowledge of vertical, complementary, and supplementary angles to find missing angles in this exercise.
https://www.khanacademy.org/math/cc-seventh-grade-math/cc-7th-geometry/cc-7th-unknown-angle-algebra/e/find-missing-angles

Angle Relationships Quiz
Test your knowledge about angle relationships with this interactive online quiz.
https://quizlet.com/152463094/test

Solving for Unknown Angles from Khan Academy
Use your knowledge of supplementary and complementary angles to solve questions of varying difficulty. Some questions involve writing and solving an equation.
https://www.khanacademy.org/math/cc-seventh-grade-math/cc-7th-geometry/cc-7th-angles/e/solving-for-unknown-angles

Angle Sums
Examine the angles in a triangle, quadrilateral, pentagon, hexagon, heptagon or octagon. Can you find a relationship between the number of sides and the sum of the interior angles?
http://illuminations.nctm.org/Activity.aspx?id=3546

Angles in a Triangle Quiz
Find the size of the angle marked with a letter in each triangle.
http://www.transum.org/software/SW/Starter_of_the_day/Students/AnglesInTriangle/Quiz.asp

CONSTRUCTIONS

Geometric Construction
These lessons cover constructions for perpendicular lines, an equilateral triangle, angle bisection, parallel lines, and copying an angle. They include explanations, interactive animations, and self-check questions.
http://www.absorblearning.com/mathematics/demo/units/KCA006.html

Animated Geometric Constructions
Simple animations that show how to do basic geometric constructions.
http://www.mathsisfun.com/geometry/constructions.html

Constructing a Triangle with Three Known Sides
A short demonstration showing how to construct a triangle with three known sides using a compass and a ruler.
http://www.mathsisfun.com/geometry/construct-ruler-compass-1.html

Constructing Triangles
Practice constructing triangles in this interactive activity from Khan Academy.
https://www.khanacademy.org/math/7th-engage-ny/engage-7th-module-6/7th-module-6-topic-b/e/constructing-triangles

Triangle Inequality Theorem
Answer questions about the third side of a triangle when the lengths of two sides are given.
https://www.khanacademy.org/math/7th-engage-ny/engage-7th-module-6/7th-module-6-topic-b/e/triangle_inequality_theorem

PI AND THE CIRCUMFERENCE OF A CIRCLE

A Rolling Circle Illustrating Pi
Two animations where a circle with diameter 1 rolls on a number line one complete roll. Of course having rolled once around its circumference, it now lands at pi.
http://www.mathwarehouse.com/animated-gifs/images/demonstration-of-PI-in-math-animation.gif

http://i.imgur.com/dsCw0.gif

Approximating Pi
How did Archimedes find the approximate value of pi? This interactive tool illustrates Archimedes' basic approach with inscribed or circumscribed polygons.
http://www.pbs.org/wgbh/nova/physics/approximating-pi.html

5 Trillion Digits of Pi
As of 2014, the world record for computing digits of pi was over 13 trillion digits.
http://www.numberworld.org/digits/Pi/

Radius, Diameter and Circumference
Practice finding the radius, diameter, or circumference of a circle in this interactive online activity.
https://www.khanacademy.org/math/geometry/hs-geo-foundations/hs-geo-area/e/radius_diameter_and_circumference

Circumference Quiz
Test your skills with this interactive self-check quiz.
https://maisonetmath.com/circumference-quizzes/302-circumference-of-circles

Amazing History of Pi
A short and simple introduction to the history of pi.
https://web.archive.org/web/20170204081925/http://ualr.edu/lasmoller/pi.html

AREA OF CIRCLE

Interactive Area of a Circle
Explore and discover the relationship between the area, radius, and graph of a circle. Just click and drag the points.
http://www.mathwarehouse.com/geometry/circle/interactive-area.php

Circle Tool from Illuminations
How do the area and circumference of a circle compare to its radius and diameter? Investigate these relationships in the Intro and Investigation sections and then hone your skills in the Problems section.
http://illuminations.nctm.org/Activity.aspx?id=3547

Area of a Circle and Its Formula
Seven practice problems concerning the area of the circle with full solutions.
http://www.mathwarehouse.com/geometry/circle/area-of-circle.php

Quiz: Area of a Circle
Practice finding the area of a circle in this interactive online quiz.
http://www.phschool.com/webcodes10/index.cfm?wcprefix=bja&wcsuffix=1003&area=view

Area and Perimeter Practice
A 10-question quiz that will let you practice finding the area and circumference of a circle.
http://www.thatquiz.org/tq-4/?-j201g-la-p1ug

Area and Circumference of a Circle - Test from BBC Bitesize
The level of difficulty increases as you proceed from the first to the last question in this 10-question quiz about the area and circumference of a circle.
http://www.bbc.co.uk/bitesize/quiz/q90581037

Compound Circles
Find the area and perimeter of a shape made from two semicircles in this interactive activity.
https://flashmaths.co.uk/viewFlash.php?id=61

Area and Circumference of a Circle
A 15-question multiple-choice quiz.
http://www.proprofs.com/quiz-school/story.php?title=area-circumference--circle

Area of a Circle
An interactive activity where you cut a circle into wedges in order to determine its area.
http://www.learner.org/courses/learningmath/measurement/session7/part_b/index.html

Area of Circles
Cut a circle into sectors and rearrange the sectors to form a figure close to a parallelogram in this interactive activity. By increasing the number of sectors, the figure gets closer and closer to a perfect parallelogram.
https://www.geogebra.org/m/fyqAUV22

Area of a Circle, How to Get the Formula
An animation that illustrates how we can find the area of a circle by drawing triangles into it. The area of the circle is then the limit of the sum of the areas of the interior triangles as the number of triangles goes to infinity.
https://www.youtube.com/watch?v=YokKp3pwVFc

AREA AND PERIMETER

Area Tool
Use this tool to determine how the length of the base and the height of a figure can be used to determine its area. How are the area formulas for trapezoids, parallelograms, and triangles similar? How do they differ?
http://illuminations.nctm.org/Activity.aspx?id=3567

Free Worksheets for the Area of Triangles, Quadrilaterals, and Polygons
Generate customizable worksheets for the area of triangles, parallelograms, trapezoids, or polygons in the coordinate grid. Options include scaling, image size, workspace, border, and more.
http://www.homeschoolmath.net/worksheets/area_triangles_polygons.php

BBC Bitesize - Area
Brief revision (review) "bites," including a few interactive questions, about the area of triangles, parallelograms, and compound shapes.
http://www.bbc.co.uk/bitesize/ks3/maths/measures/area/revision/4/

Geometry Area/Perimeter Quiz from ThatQuiz.org
Find the area and perimeter of rectangles, triangles, parallelograms, and trapezoids. You can modify the quiz parameters, for example to omit a shape, or to solve for an unknown side when perimeter/area is given.
http://www.thatquiz.org/tq-4/?-j1i00f-lc-p0

Find Areas of Trapezoids
Practice finding the areas of trapezoids in this interactive online activity from Khan Academy.
https://www.khanacademy.org/math/cc-sixth-grade-math/cc-6th-geometry-topic/cc-6th-area/e/areas_of_trapezoids_rhombi_and_kites

Find Areas of Compound Shapes
Practice finding the areas of complex shapes that are composed of smaller shapes in this interactive online exercise.
https://www.khanacademy.org/math/cc-sixth-grade-math/cc-6th-geometry-topic/cc-6th-area/e/area-of-quadrilaterals-and-polygons

Area of Composite Shapes
Find the areas of combined shapes made up of one or more simple polygons and circles.
http://www.transum.org/software/SW/Starter_of_the_day/Students/Areas_of_Composite_Shapes.asp?Level=4

Metric Area
Read a short, illustrated lesson about metric area. Then, click on the questions at the bottom of the page to practice.
http://www.mathsisfun.com/measure/metric-area.html

Metric Unit Conversions: Area
Use these printable worksheets for review or to brush up on metric conversion skills.
http://www.dadsworksheets.com/worksheets/metric-si-unit-conversions-metric-si-area.html

Volume of Prisms and Cylinders
https://www.proprofs.com/quiz-school/story.php?title=MTA2MzA07UGR

Angle Relationships

A **ray** has a starting point and continues indefinitely in one direction (indicated by one arrowhead).

An **angle** consists of **two rays that start at the same point**, called the **vertex**. Each ray is called a **side** of the angle.

We can denote the angle on the right as angle BAC, or using the symbol "∠" for "angle," as ∠BAC.

Note that we list the vertex point in the middle: it is ∠B<u>A</u>C, not ∠ABC. We could also name it ∠CAB.

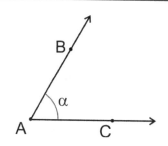

In mathematics, we also often denote angles with the beginning letters of the Greek alphabet: α (alpha), β (beta), γ (gamma), and δ (delta). So ∠BAC can also be called "angle α."

Two angles are **adjacent** if they have a **common vertex and share one side.**

In the image on the right, ∠α and ∠β are adjacent (side-by-side) angles.

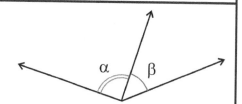

1. How many angles do you see in the picture? _____

    How many degrees do these angles measure?

    ∠ABC = _____°

    ∠CBD = _____°

    ∠ABD = _____°

    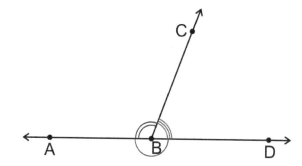

    What is the sum of ∠ABC and ∠CBD? _____°

    What is the sum of all three angles? _____°

2. Measure the angles. Calculate their sum.

    ∠A = _____°

    ∠B = _____°

    ∠C = _____°

    ∠D = _____°

    Sum of the angles = _____°

    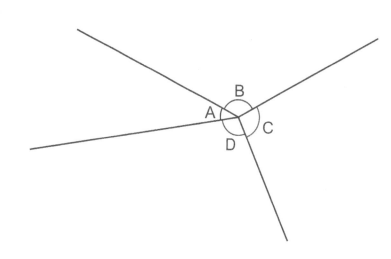

The angles ∠α and ∠β in this image are adjacent, and they form a straight angle (an angle of 180 degrees). They are called **supplementary angles**.

Two angles are supplementary if their **sum is 180 degrees**:

$$\angle\alpha + \angle\beta = 180°$$

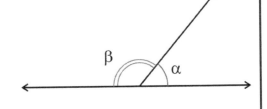

The angles ∠α and ∠β in this image are adjacent, and they form a right angle. They are called **complementary angles**.

Two angles are complementary if their **sum is 90 degrees**:

$$\angle\alpha + \angle\beta = 90°$$

We can also say, "α complements β."

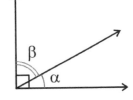

Here's a mnemonic to help you remember the difference: <u>S</u>upplementary angles form a <u>S</u>traight line, and <u>C</u>omplementary angles form a <u>C</u>orner (a right angle).

Supplementary angles don't have to be adjacent, and neither do complementary angles.

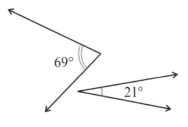

These are still complementary angles, because 21° + 69° = 90°.

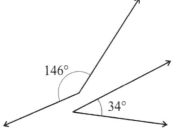

These are still supplementary angles, because 146° + 34° = 180°.

3. **a.** Draw a 38° angle. Then draw an adjacent angle that complements it.

   b. Draw an 82° angle. Then draw an adjacent angle that supplements it.

114

4. Write an equation for each of the unknown angles. Then solve it. Do not measure any angles.

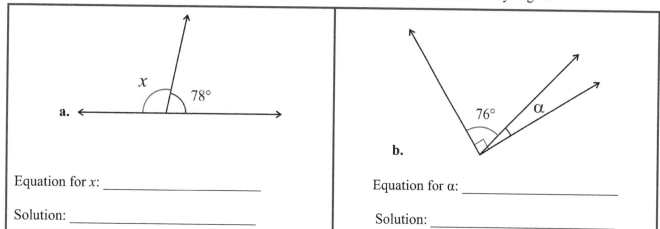

a.

Equation for x: _____

Solution: _____

b.

Equation for α: _____

Solution: _____

5. Figure out the missing entries in the table without actually measuring any angles. Remember that a full circle is 360°.

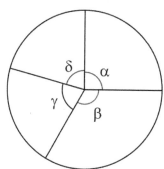

Angle	Degrees	Fraction	Percentage
α		1/4	
β	120°		
γ			
δ	75°		

6. Figure out the missing entries in the table without actually measuring any angles.

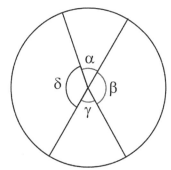

Angle	Degrees	Fraction	Percentage
α	50°		
β			
γ		1/6	
δ			

7. What is the value of $x + y$?

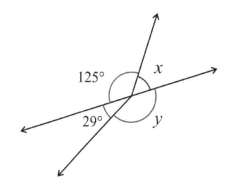

115

Vertical angles

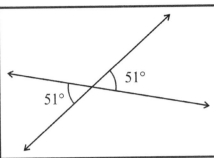

When two lines intersect, they form four angles. The two opposite angles are called **vertical angles**.

Vertical angles are **congruent**. (They have the same angle measure.)

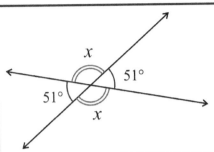

Example 1. In this picture, you see two pairs of vertical angles. What is the measure of angle *x*?

Notice that the angle *x* and the 51° angle are supplementary (they form a straight line), which means that $x + 51° = 180°$. From that we get $x = 180° - 51° = 129°$.

8. Find the measures of angles α and β without measuring.

∠α = _____ °   ∠β = _____ °

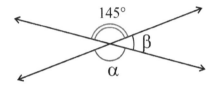

9. **a.** Find a pair of vertical angles in the figure.

b. Write an equation for α, and solve it.
   Hint: Look for angles that form a straight line.

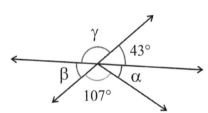

c. What is the measure of ∠γ?

10. Calculate the measures of angles α, β, γ, and δ.
   Hint: Look for vertical angles and for angles that form a straight line.

∠α = _____ °   ∠β = _____ °

∠γ = _____ °   ∠δ = _____ °

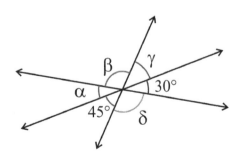

11. In this figure, lines *k* and *m* are parallel and line *l* intersects them both.

    a. Mark all the pairs of vertical angles in the figure.

    b. Measure or calculate all the eight angles. Mark them in the figure.

    What do you notice?

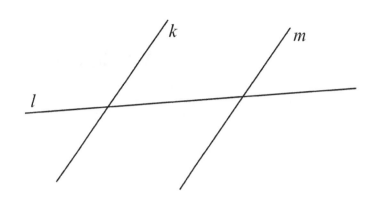

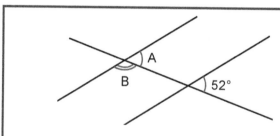

Lines *l* and *k* are parallel, and line *m* intersects them both.

We call angles A and C **corresponding angles** because they are oriented in the same sense in matching "corners."

Angles B and D are also corresponding angles.

Because the rays that form them are parallel, **corresponding angles are equal.**

12. One angle is given. Find the measures of the marked angles without measuring.

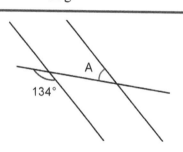

a. ∠A = _____ °   ∠B = _____ °

b. ∠A = _____ °

13. a. This figure has *two* pairs of parallel lines, and angle A = 109°. Reason out the measures of the other angles.

∠B = _____ °   ∠C = _____ °   ∠D = _____ °

b. What familiar polygon is formed in the middle?

14. Reason out the measures of angles A, B, and C in this parallelogram.

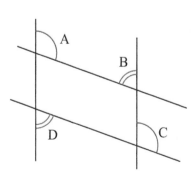

117

Angles in a Triangle

1. Using a ruler, draw any triangle in this space or in your notebook. Measure all its angles. Calculate their sum.

   The sum of the angles is _____°.

2. Draw a different triangle. Measure all its angles. Calculate their sum.

   The sum of the angles is _____°.

You have probably already made a guess that **the sum of the angles in a triangle is 180°**. That is true. Here is a **proof** for it. "Proof" means that we use already established principles to show that some new statement is true.

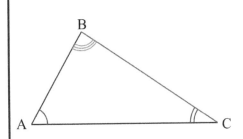

Let ABC be a triangle (a generic triangle we will use in the proof).

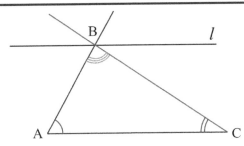

We draw line l so that it is parallel to AC and passes through point B. We also extend AB and CB as shown in the figure.

Extending the rays forms the adjacent angles α', β, and γ'.

Because α and α' are corresponding angles, $\angle \alpha' = \angle \alpha$.

Because β and β' are vertical angles, $\angle \beta' = \angle \beta$.

And because γ and γ' are corresponding angles, $\angle \gamma' = \angle \gamma$.

Therefore, the sum $\angle \alpha' + \angle \beta' + \angle \gamma'$ is equal to the sum $\angle \alpha + \angle \beta + \angle \gamma$.

Since l is a straight line, the three angles α', β', and γ' form a straight angle, so the sum of their measures is 180°. In the previous step we just demonstrated that $\angle \alpha' + \angle \beta' + \angle \gamma' = \angle \alpha + \angle \beta + \angle \gamma$, so, since $\angle \alpha' + \angle \beta' + \angle \gamma' = 180°$, this means that $\angle \alpha + \angle \beta + \angle \gamma = 180°$, too (which is what we wanted to prove).

3. In each problem, write an equation for the unknown angle and solve it. *Do not measure.*

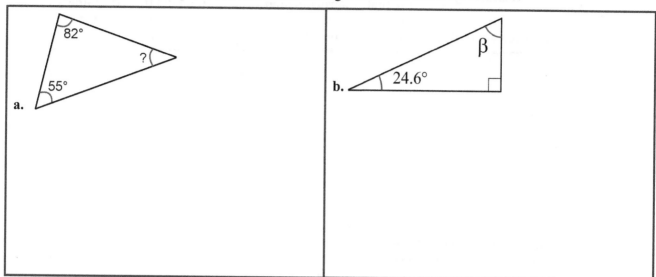

4. **a.** A certain triangle has three equal angles. What is the measure of each angle? _____°

   b. Draw one such triangle using your protractor.
   Make each of its sides 5 cm long.

   c. This triangle has a special name.
   What is it called?

5. What is the measure of the top angle?

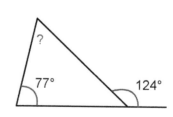

Let's review how to draw a line that is perpendicular (at a right angle) to a given line
First draw a point (a dot) on the given line. For the next step you will need a protractor.

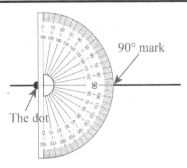

Align the point and the straight side of your protractor. Also align the given line and the 90° mark on the protractor.

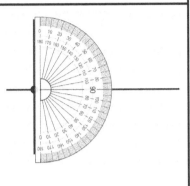

Draw the line.

6. **a.** The figure below on the right shows line segment $\overline{AB}$ that is 4 cm long. Draw another line segment, $\overline{AC}$, from A that is 6 cm long and perpendicular to $\overline{AB}$. Then draw the segment $\overline{CB}$.

   b. What kind of triangle is formed?

   c. Are any two angles in your triangle complementary? If so, which ones?

   d. Are any two angles in your triangle supplementary? If so, which ones?

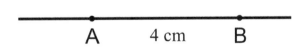

7. Can you draw a triangle that has two obtuse angles? Why or why not?

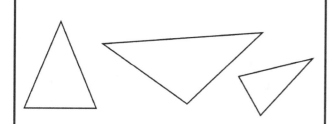

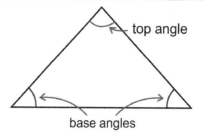

An **isosceles triangle** has two sides that are equal. Think of it as a "same-legged" triangle, the "legs" being the two sides that are the same length.

Not only does an isosceles triangle have two sides that are equal, but two of its angles are also equal. They are called the **base angles.** The remaining angle is called the **top angle**.

8. The top angle of an isosceles triangle measures 25°. What do the other two angles measure?

9. Draw an isosceles triangle with 40° base angles and a 4 1/2-inch side between the two angles.

10. Draw an isosceles triangle with a 64° top angle and two 4.3-cm sides. What is the measure of the base angles?

Classification of triangles

Every triangle is

- acute, obtuse, or right (classified according to angles), and
- equilateral, isosceles, or scalene (classified according to sides).

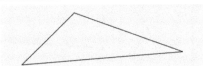

This is an obtuse scalene triangle.

11. **a.** Draw a triangle with 65° and 50° angles and a 7.5-cm side between those two angles. Start out by drawing the 7.5-cm side near the bottom of the drawing space below.

   b. *Calculate* the third angle. It is _____°. Then measure it in your triangle to check.

   c. Classify your triangle according to its sides and angles:

   It is _____ and _____.

12. Lines *l* and *m* are parallel. Figure out the measure of the unknown angle.

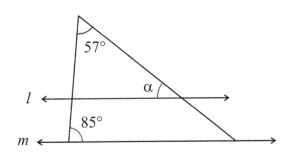

Any quadrilateral can be divided into two triangles by drawing a diagonal. We can use this idea to easily prove what the sum of the angles of any quadrilateral must be.

a. What is the sum of the angles of any quadrilateral?

b. Prove it.

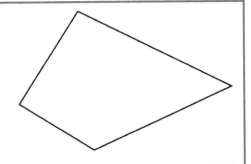

Puzzle Corner

Basic Geometric Constructions

Geometric constructions are drawings done using only these two tools:
- a compass
- a straightedge (a ruler).

A compass allows you to draw **points that are at a specified distance from a certain point** (a circle's center point). This fact proves out to be very useful in geometric drawings!

A **straightedge** is a ruler without measurement units (such as cm or in). It is used only to draw straight lines. You can use your normal ruler. Just ignore the units of measurement on it.

You will complete most of the exercises of this lesson using <u>only a compass and a straightedge</u> or drawing software. All you need is the ability to draw circles from their center point and to draw straight lines, so even the drawing tools in a word processor program are sufficient.

Tips: 1. In MS Word, go to View → Toolbars → Drawing to see the drawing tools.
2. In many programs, holding the Control and Shift keys while drawing a circle forces the circle to be drawn as a perfect circle (not as an ellipse) and from its center point (not from the side).

Copy a Line Segment
Our task is to draw a copy of a given line segment, or in other words to draw another line segment of the same length, anywhere on the paper. Start out by drawing a long line and drawing a point on it (A'). Now, think: how can you use the <u>compass</u> to find where the point B' should be so that $\overline{A'B'}$ is as long as $\overline{AB}$?

1. Copy the line segment.

2. Draw a line segment that is as long as these two line segments together.

123

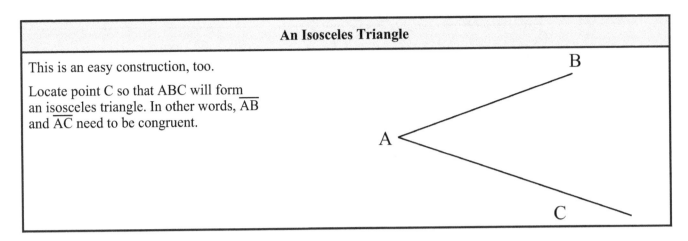

3. Draw any isosceles triangle on blank paper. Also draw one with drawing software.
   Hint: Start out by drawing any angle.

4. Draw an isosceles triangle with two sides this long: ─────────────────

An Equilateral Triangle

An equilateral triangle has three congruent sides. Its name helps you remember what it means: "equi" refers to equal, and "lateral" to sides, so "equi-lateral" refers to an equal-sided triangle.

This means its vertices are at the same distance from each other. Keep in mind that a compass helps us find points that are at the same distance from each other!

A equilateral triangle has another special feature also: each of its angles is 60 degrees.

The line segment AB marks the side of the triangle. Draw a circle using point A as the center point and $\overline{AB}$ as the radius. The third vertex of the triangle MUST lie on this circle... because its distance to B is equal to $\overline{AB}$.	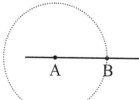
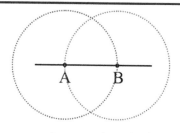 Can you see what was done in this picture?	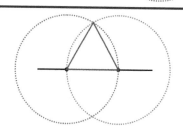 The triangle is done!

5. Draw an equilateral triangle using this line segment as the base.

6. **a.** Draw any equilateral triangle on blank paper. You can choose how long the sides are.

   b. Draw another equilateral triangle with drawing software.

A Triangle with Three Given Sides

Our task is to construct a triangle using these line segments as its sides. We are essentially given three *distances*, which means a compass will help.

Start out by choosing any of the line segments as the base. Copy that line segment.

Draw a circle using one of the end points of the first line segment as the center point and the *second* line segment as the radius.

Then draw a circle using the other end point of the first line segment as the center point and the *third* line segment as the radius.

Where is the third vertex of the triangle?
Lastly, draw the two sides of the triangle.

7. Draw a triangle using these three line segments as sides.

8. **a.** Draw a triangle using these three line segments as sides.

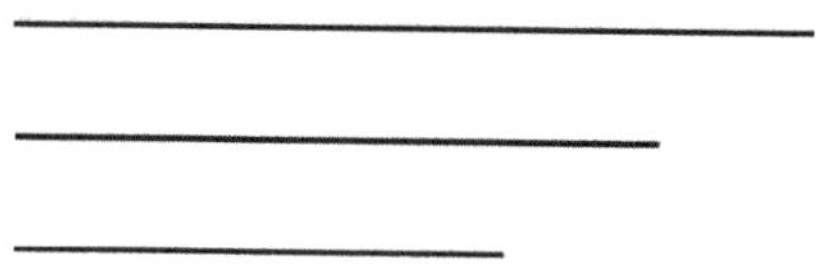

   b. Classify the triangle according to its angles and sides.

   _____

   _____

   _____

9. Draw a triangle with sides 4.5 cm, 6.8 cm, and 5.7 cm long. This time, you will need a regular centimeter-ruler and a compass.

10. **a.** The table lists three sets of lengths. If these are used as lengths of sides for a triangle, one of them does not make a triangle. Which one? (Try to draw the triangles on a blank paper.)

8 cm, 6 cm, 10 cm	3 cm, 12 cm, 8 cm	10 cm, 13 cm, 15 cm

    b. Change one of the lengths in the set that didn't make a triangle so that the three lengths will form a triangle.

Triangle Inequality

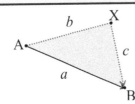

Which way is shorter? From A to B, or from A to B via X?

It is always shorter to go straight from A to B (distance a) than to first go from A to X (distance b) and then from X to B (distance c).

In symbols, $a < b + c$.

The triangle inequality also gives us a way to determine if three lengths can form a triangle.

Three lengths a, b, and c form a triangle if — and only if — the sum of any two is greater than the third.

Otherwise you would have a triangle where traveling along two of the sides would be a shorter distance than traveling along the third side.

For example, the lengths 2 cm, 2 cm, and 5 cm cannot form a triangle, because 2 + 2 is not greater than 5.

In symbols, for a, b, and c to form a triangle, each of the following three inequalities must be true:

- $a + b > c$
- $a + c > b$
- $b + c > a$

Can you see why this triangle is fake?

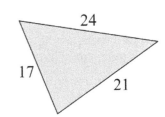

11. Write the three triangle inequalities, $a + b > c$, $a + c > b$, and $b + c > a$ for this triangle.

  _____ + _____ > _____

  _____ + _____ > _____

  _____ + _____ > _____

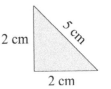

12. Which sets of lengths do not make a triangle?

| 7 in, 3 in, 2 in | 10 cm, 13 cm, 17 cm | 6 yd, 8 yd, 11 yd | 7 m, 10 m, 2 m |

13. Fill in: In a triangle with sides 50 cm and 65 cm, the third side must be at least _____ cm.

Puzzle Corner

Let a, b, and c be the sides of a triangle. According to the triangle inequality, $a < b + c$. But what if we allow equality so that $a = b + c$? Use an actual numerical example and a drawing to explain what happens in that case.

P.S. Mathematicians do actually allow for equality in the triangle inequality and write it as $a \leq b + c$.

More Constructions

(This lesson is optional)

1. The instructions below explain how to construct a perpendicular line through a point on a line. The last step is missing, though. Finish the construction.

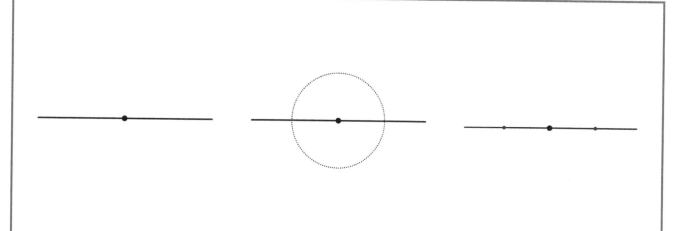

(1) You are given a line and a point on it. The task is to draw a perpendicular line through this point.

(2) First, draw any circle using the given point as a center. Mark the points where your circle intersects the line. Then you can erase the circle.

(3) Now you have two helping points on both sides of the given point. Complete the construction. Think back to the constructions you have already learned!

2. The instructions below explain how to construct a perpendicular line through a point *not* on the line. The last step is missing, though. Finish the construction.

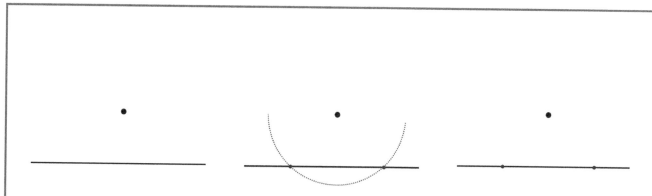

(1) You are given a line and a point *not* on it. The task is to draw a line, through the point, that is perpendicular to the original line.

(2) First, draw any circle using the given point as the center. Mark the points where your circle intersects the line. Then you can erase the circle.

(3) Now you have two helping points on both sides of the given point. Complete the construction. Think back to the constructions you have already learned!

The following problems let you practice the constructions explained above and also discover new ones. You can draw them on separate paper if you need to. Keep thinking. That's part of the fun of drawing. Just memorizing how constructions are done is not as valuable as thinking for yourself and finding a way they can be done! Also remember that the same construction can often be done in several different ways.

3. Draw any line. Then draw a line perpendicular to it.

4. Draw a rectangle that is not a square. (Hint: start out by drawing a long line, and mark the two vertices of the rectangle with dots somewhere on that line.)

5. **a.** Construct a perpendicular line from the vertex A to the opposite base.

   b. Choose another vertex, and construct a perpendicular line from that vertex to the opposite base.

   c. Your two lines meet at a point inside the pentagon. Using that as a center point, draw a circle so it passes through A and all the other vertices. You have just circumscribed the pentagon!

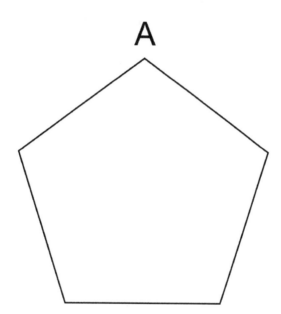

6. The instructions below explain how to draw an altitude to a triangle. The last step is missing, though. Finish the construction.

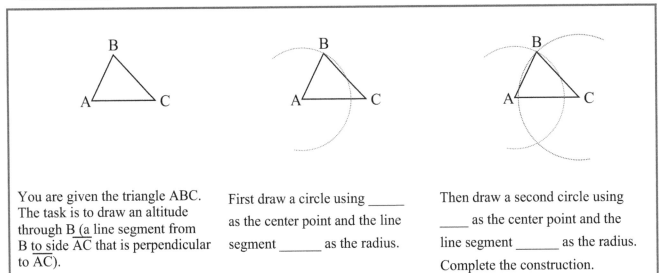

You are given the triangle ABC. The task is to draw an altitude through B (a line segment from B to side $\overline{AC}$ that is perpendicular to $\overline{AC}$).

First draw a circle using _____ as the center point and the line segment _____ as the radius.

Then draw a second circle using _____ as the center point and the line segment _____ as the radius. Complete the construction.

7. Draw an altitude to each triangle from the top vertex. Notice the second triangle is obtuse, so the altitude will be outside of the triangle.

8. Draw all three altitudes to this triangle. What special thing do you notice?

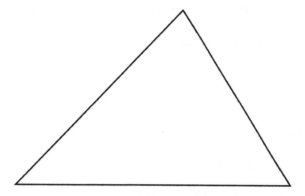

Drawing Problems

A triangle can be described using its angles and its sides. This of course gives us six pieces of information: the measures of three angles and the lengths of three sides.

However, we don't actually need all six items to determine a triangle. We can get by with less information. Let's find out how many of the items it takes to determine a unique triangle.

Example 1. Draw a triangle with a 40-degree angle and one 5-cm side.

Clearly, we can draw triangles of several different shapes that fit this information. These two pieces of information (one angle and one side) are not enough to uniquely determine a triangle. We don't even know if the angle is opposite to the 5-cm side or next to it.

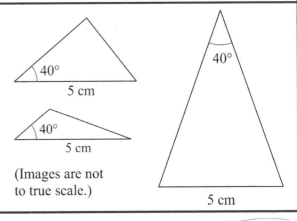

(Images are not to true scale.)

Example 2. Draw a triangle with sides 7 cm, 8 cm, and 9 cm long. Can you draw several of different shapes?

We already studied how to draw a triangle with three given sides using a compass and straightedge. That process produces a unique triangle. If you draw it again, you will get a triangle that is an exact copy of the first one, maybe rotated, flipped, or in a different place on your paper, but the two are **congruent**: if one were placed on top of the other, they would match exactly.

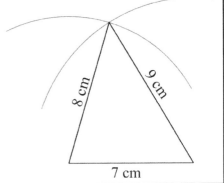

1. Does the information given define a unique triangle? If yes, say so, and draw the triangle. If not, prove that it doesn't by drawing at least two non-congruent (different-shaped) triangles that satisfy the given conditions.

   a. A triangle with two 60-degree angles

   b. An isosceles triangle with two 7-cm sides

   c. An isosceles triangle with two 7-cm sides and a 12-cm base

   d. An equilateral triangle with 6 1/2-inch sides

> **In a parallelogram, the opposite angles are equal.**
>
> Not only that, but any two neighboring angles sum to 180 degrees:
> ∠A + ∠B = 180°.
>
> Of course, the sum of all four angles is 360°, just like in any quadrilateral.

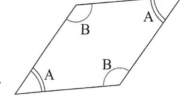

2. **a.** Draw a parallelogram with one 65° angle and sides that are 8 cm and 5 cm long.
   Hint: Start out by drawing the 65° angle. Then draw the base side.

   b. Are any of the angles in the parallelogram complementary?
   If so, which ones?

   c. Are any of the angles in the parallelogram supplementary?
   If so, which ones?

3. A parallelogram has two 50° and two 130° angles. Does this information define a unique parallelogram?
   If so, draw it. If not, draw several non-congruent parallelograms that fit this condition.

135

4. One angle of a rhombus (a parallelogram with four equal sides) is 115° and its side measures 2.5 inches.

   a. Calculate the angle measures of the other three angles: they measure _____°, _____° and _____°.

   b. Does the information given determine a unique rhombus?

   c. If so, draw the rhombus. If not, draw several different rhombi that fit the description.

5. A parallelogram has 3-in and 4.5-inch sides and a 60°-angle between them. Does this information define a unique parallelogram? If so, draw it. If not, draw several non-congruent (different-shaped) parallelograms that fit this condition.

6. Carpenters often use the so-called 3-4-5 triangle, which is a triangle with sides 3, 4, and 5 units long. The units can be inches, centimeters, meters, or even other sizes, such as 10-cm or 3-in. "units." For example, if the "unit" is 3 inches, then the triangle's sides become 9 inches, 12 inches, and 15 inches.

   a. Draw at least three 3-4-5 triangles using different "units."

   b. Measure the angles of your triangles. What do you notice?

7. This cross figure is drawn here at the scale 1:5 (which means that, in reality, it is five times as big).

   a. Calculate the area of the figure in reality.
   Hint: measure the sides of the figure in centimeters.

   b. Make a new scale drawing of the original figure at the scale 1 : 6.
   (Remember, you do not see the original figure here. The original is much larger than the scale drawing below, which is at the scale 1 : 5.)

Scale 1 : 5

8. Below you see a scale drawing of a triangle, drawn at the scale 1 cm = 30 cm. Make a new scale drawing of the original figure, this time using the scale 1 : 20.

Scale 1 cm = 30 cm

9. Let's study some more about triangles. Does the information given define a unique triangle? If it does, say so, and draw the triangle. If not, prove that it doesn't by sketching at least two non-congruent (different-shaped) triangles that satisfy the given conditions.

   a. An isosceles triangle with a top angle of 90°.

   b. An isosceles triangle with a 90° top angle and a 12-cm base.

   c. A triangle with angles 50°, 40°, and 90°.

   d. A triangle with 30° and 90° angles plus a 6-cm side.

   e. A triangle with 30° and 90° angles plus a 6-cm side *between* those angles.

   f. A triangle with sides 10 cm and 8 cm long that form a 70° angle.

   g. A triangle with sides 10 cm and 8 cm long and a 70° angle (location not specified).

10. We haven't studied all the possible combinations of pieces of information (angles and sides) that determine a triangle, but we have looked at several. Based on the exercises in this lesson and logical thinking, fill in the table. You will explore this topic more in high school geometry.

Givens	Determines a unique triangle? (yes/no)
Three sides	
Two sides and a given angle formed by those sides	
Two sides and a given angle (location of angle not specified)	
Three angles	
Two angles and a side between them	
Two angles and a side (location of side not specified)	
One side and one angle	

11. The image illustrates the principle for finding the sum of the angles of a pentagon.

    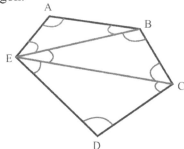

    a. Figure out what the principle is and fill in:

    The sum of the angles of a pentagon is _____°.

    b. A regular pentagon has five equal angles and five equal sides.

    Each angle in a regular pentagon must measure _____°.

    c. Your friend has drawn a particular regular pentagon on paper. Without seeing it, which information would you need to know about her figure in order to be able to draw an identical copy of her pentagon?

---

a. A rhombus is a quadrilateral with four equal sides. Keeping in mind that a compass allows you to draw points that are at a specified distance from the center point, devise a way to construct a rhombus using a compass and straightedge only. You decide how long each side will be. Use blank paper or a notebook.

b. Construct a regular hexagon. Start out by constructing two equilateral triangles using the usual construction for an equilateral triangle with two circles (see the image). The hexagon will be inside one of these circles.

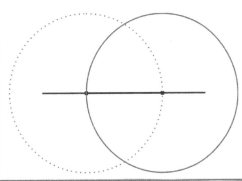

Circumference of a Circle

- The **circumference** of a circle is the perimeter, or outside curve, of the circle.
- The **diameter** is any line segment from circumference to circumference that goes through the center point of the circle.
- The **radius** is any line segment from the circumference to the center point. The radius is exactly half of the diameter. It is what you use to draw a circle with a compass.

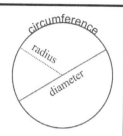

There exists an amazing relationship between the circumference and the diameter of every circle! In this lesson we will study this relationship.

You may use a calculator for every problem in this lesson.

1. Find at least five circular objects, such as a plate, a can, a glass, and so on. Measure the diameter of each circle with a ruler. Measure the circumference of each circle by placing a string around the object, and then measuring the length of the string. Record your results in the table.

Object	Circumference	Diameter	Circumference ÷ diameter

In the last column, divide the circumference by the diameter using a calculator (separately for each object). In other words, calculate the ratio of the circumference to the diameter. If you have measured accurately, you should get a number that is <u>a little over 3</u> in the last column.

> Even thousands of years ago, people knew that the circumference of a circle was *about* three times its diameter.
>
> Ancient Egyptians used the number 22/7 or 3 1/7 (which is about 3.14285) instead of 3. If the diameter of a circle was 2 units long, the Egyptians would have calculated the circumference to be (22/7) · 2 = 44/7 = 6 2/7 units.

2. Find the circumference of each circle by multiplying its diameter by 22/7. Round the answers to one decimal.

a. Diameter 5 cm	**b.** Radius 3 in.	**c.** Diameter 3 m
Circumference =	Circumference =	Circumference =

141

In reality, the ratio of a circle's circumference to its diameter is neither exactly 3 nor 22/7. It is **pi**: a number that is about 3.1416 and denoted by the Greek letter π. The value of π has been calculated to many millions of decimal digits. Here are some of the first digits:

π ≈ 3.14159265358979323846264338327950288419716939937510582097494459230781 6406286...

The decimal expansion of π goes on forever without ever having any pattern in the digits! This means it is an **irrational number**: you cannot write it as a fraction *a/b*, where *a* and *b* are integers.

In calculations, you can use π ≈ 3.14, π ≈ 22/7, or the π-button on your calculator (if you have it).

3. A circle is drawn on the ground. Your 15-foot jump rope is just enough to go around it. Match the approximate measurements and the terms.

diameter	about 15 ft
radius	about 2.5 ft
circumference	about 5 ft

4. Find the circumference of these circles. Use π ≈ 3.14. Give your answer to the same decimal accuracy as the dimension given in the problem.

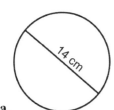

a.

C = _____

b.

C = _____

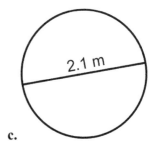

c.

C = _____

5. **a.** Draw a circle with a diameter of 5 cm.
   Use a compass.
   What is its radius?
   And its circumference?

   b. Draw a *square* that has the same
   perimeter as your circle.
   How long is the side of your square,
   to the nearest millimeter?

As a formula, we express this relationship as $\dfrac{C}{d} = \pi$.

This means that when you divide the circumference (C) by the diameter (d), you get pi. That's exactly what you did in exercise 1.

Multiply both sides of that formula by *d* to get another formula for the same relationship: $C = \pi d$. It allows us to calculate the circumference when the diameter is known.

Example 1. The diameter of a circle is 4.5 m. What is its circumference?

We multiply the diameter by π to get C = π · 4.5 m ≈ 3.14 · 4.5 m ≈ 14.1 m.

Example 2. The circumference of a circle is 30 ft. What is its diameter?

Notice this is the *opposite* problem to example 1. Therefore, we do the opposite and *divide* the circumference by π to get $d = \dfrac{C}{\pi} \approx \dfrac{30 \text{ ft}}{3.14} \approx 9.6$ ft.

6. **a.** The diameter of a circle is 5.60 km. What is its circumference? Its radius?

   b. The circumference of a circle is 120 cm. What is its diameter? Its radius?

7. Fill in the table. Use π ≈ 3.14 and a calculator. Round to one decimal digit.

	Circle A	Circle B	Circle C	Circle D
Circumference	14 cm			7.5 in
Diameter		2.5 in		
Radius			8.3 m	

8. **a.** This shape consists of two half-circles and a rectangle. The side of each little square in the grid is 1 cm long. Find the perimeter of the shape to the nearest centimeter.

   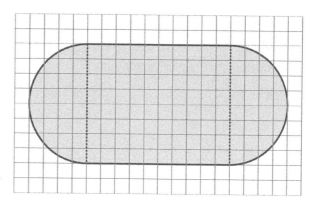

   b. If each little square on the grid measured 50 cm instead of 1 cm, what would the perimeter of the shape be?

Area of a Circle

The area of a circle is given by this formula: $A = \pi r^2$, where r is the radius of the circle.

Read the formula as: "Area equals pi r squared." It means that you first multiply the radius by itself and then multiply the result by π.

You can remember the formula by thinking, "Pie are square." Of course pies are usually round, not square! It is bad English, as well, but the purpose of this silly mnemonic is just to help you remember the formula.

Example 1. The radius of a circle measures 8 cm. What is its area?

We use the formula: $A = \pi r^2 = \pi \cdot 8 \text{ cm} \cdot 8 \text{ cm} \approx 3.14 \cdot 64 \text{ cm}^2 = 200.96 \text{ cm}^2$, or about 200 cm².

Remember to always give your answer for an area in <u>square</u> units, be it square inches, square centimeters, square meters, square feet, *etc*. If no measuring unit is given, use "square units."

You can use a calculator for all the problems in this lesson.

1. Estimate the area of the circles by counting squares and parts of squares. After that, calculate the area to the nearest tenth of a square unit.

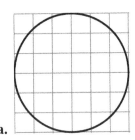
a.

Estimation: _____ square units

Calculation: _____ square units

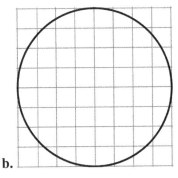
b.

Estimation: _____ square units

Calculation: _____ square units

2. Find the areas of these circles.

a. A circle with a radius of 7.0 cm. Round the answer to the nearest ten square centimeters (to 2 significant digits). Area =	**b.** A circle with a radius of 10 1/4 in. Round the answer to the nearest square inch (to 3 significant digits). Area =
c. A circle with a *diameter* of 75.0 cm. Round the answer to the nearest ten square centimeters (to 3 significant digits). Area =	**d.** A circle with a radius of 17 ft 4 in. Round the answer to the nearest thousand square inches (to 3 significant digits). Area =

3. **a.** Find the area of the circle to the nearest tenth of a square unit.

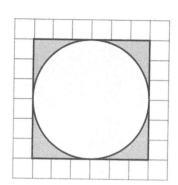

   b. Find the area of the shaded figure to the nearest tenth of a square unit.

4. A square is drawn inside a circle. The diameter of the circle is 15.0 cm, and the side of the square is 10.6 cm.

   a. Find the area of the circle to the nearest ten square centimeters.

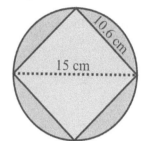

   b. Find the area of the square to the nearest square centimeter.

   c. What percentage of the area of the circle is the area of the square?

   Note: For this calculation, you'll need a more exact value for π than 3.14, so use 3.1416 or the π-button on your calculator. Also, if you use the answers from (a) and (b), use the exact (not rounded) answers.

5. The radius of a certain circle is 8 units. Which expression can you use to calculate the circumference of that circle?

   a. $\pi \cdot 16$ **b.** $\pi \cdot 8$ **c.** $\dfrac{8}{\pi}$ **d.** $\dfrac{\pi}{8}$

6. The circumference of a certain circle is 14 units. Which expression can you use to calculate the diameter of that circle?

   a. $\pi \cdot 14^2$ **b.** $\pi \cdot 14$ **c.** $\dfrac{14}{\pi}$ **d.** $\dfrac{\pi}{14}$

7. The diameter of a certain circle is 6 units. Which expression can you use to calculate the area of that circle?

   a. $\pi \cdot 6^2$ **b.** $\pi \cdot 6$ **c.** $\pi \cdot 12$ **d.** $\pi \cdot 3^2$

8. Joe's pizzeria offers the following pizzas. Let's find out which is the best buy.

Pizza diameter	Cost	Area (sq. in.)	Cost per square inch ($)
10"	$7.99		
12"	$9.99		
14"	$12.99		
18"	$15.99		

a. Find the area of each pizza to the nearest tenth of a square inch.

b. Find the cost per square inch, which is the cost divided by the area, to the tenth of a cent (three decimals).

c. Which pizza is the cheapest per square inch?

d. Which gives you more to eat, two 10" pizzas or one 14" pizza?

9. a. This shape consists of two half-circles and a rectangle. The side of each little square in the grid is 1 cm long. Find the area of the shape to the nearest square centimeter.

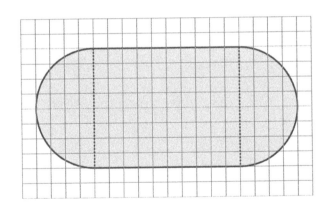

b. If the side of each little square measured 5 m instead of 1 cm, what would be the area of the shape (to the nearest ten square meters)?

Puzzle Corner

Joe calculated the area of a certain circle, using $\pi = 3.14$. The area of his circle was 153.86 cm². What was the radius of the circle?

Proving the Formula for the Area of a Circle

We will now study an informal proof for the familiar formula $A = \pi r^2$. In the first part of the proof we prove that the area of the circle is $\frac{1}{2}C \cdot r$. In part 2 we will show that $\frac{1}{2}C \cdot r$ is equal to the familiar πr^2.

Proof, part 1

Let's divide a circle into equal sectors, in this case into 12.

We can rearrange those sectors to form a figure that is very close to a parallelogram. Obviously, the area of this new shape is equal to the area of the circle. But since the area of this shape is very close to the area of a parallelogram, we can approximate its area with the formula for the area of the parallelogram.

The <u>base</u> of this parallelogram is approximately half of the circumference, or $\frac{1}{2}C$.

Why? The sectors alternate their orientations to put half of the original radius on the top of the parallelogram and half on the bottom

The <u>height</u> of the parallelogram is the radius of the circle, r (see the image at the right).

The area of a parallelogram is its base times its height, so the area of this parallelogram is

$$A = \tfrac{1}{2}C \cdot r$$

By dividing the circle into a larger number of sectors, the shape made of the sectors would get closer to a parallelogram. In fact, we can get as close to a parallelogram as we want by dividing the circle into a very large number of sectors. So the area of the circle is $A = \frac{1}{2}C \cdot r$.

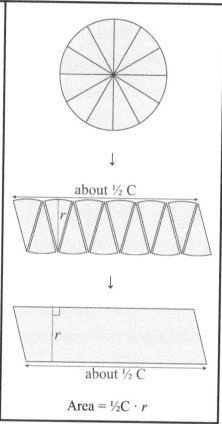

We have called this an "informal" proof because we have omitted a very important point: In order to prove that the area of the figure that looks like a parallelogram does indeed approach the area of a parallelogram as we increase the number of sectors, we need to use the concept of a **limit** from calculus.

1. **a.** Study the proof above enough that you can explain it to another person.

   b. Draw a large circle on paper. Divide it carefully and as exact as you can into 8 or 12 equal sectors and cut the pieces out. Then explain the proof you just studied to a fellow student, friend, or your teacher.

2. Let's explore how much closer to a parallelogram you can get by dividing the circle into more sectors. You can do this exercise either in drawing software or on paper. Draw a large circle, and divide it carefully into 16 or 20 equal sectors (if drawing on paper, use a protractor). Arrange the sectors into a "parallelogram" shape. Compare your shape to the shape in this lesson, which was made from 12 sectors.

3. Estimate in your head (without a calculator) the area of a circle with a radius of 5.0 ft and circumference of 31.0 ft. Use $A = \frac{1}{2}C \cdot r$.

Proof, part 2

Let's use some algebra to transform our equation $A = \frac{1}{2}C \cdot r$ into the familiar equation for the area of a circle, $A = \pi r^2$.

First, since we know that the circumference of a circle is $C = \pi d$, where d is the diameter, we can substitute πd in place of C to get

$$A = \tfrac{1}{2}C \cdot r = \tfrac{1}{2}\pi d \cdot r$$

In our final expression we want the radius instead of the diameter, so let's substitute $d = 2r$ to convert the diameter into radius:

$$A = \tfrac{1}{2}\pi d \cdot r = \tfrac{1}{2}\pi \cdot 2r \cdot r$$

Lastly, the 1/2 and 2 cancel each other, and we are left with

$$A = \pi \cdot r \cdot r = \pi r^2$$

---

4. **a.** Estimate in your head (without a calculator) the area of a circle with a diameter of 12.0 m. Use $A = \frac{1}{2}C \cdot r$ and the fact that the circumference of a circle is a little over 3 times its diameter.

   b. Use a calculator to find the exact area of this circle with the formula $A = \pi r^2$. Compare the result to your estimate from part (a).

5. Your jump rope is 10 ft long. You form it into a circle. What is the area of your circle? Use a calculator.

Area and Perimeter Problems

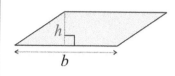
The area of a parallelogram is
$$A = bh$$
where b is the base and h is the altitude (height).

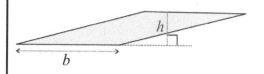
The **altitude** of a parallelogram is a perpendicular line segment from the base, or the extension of the base, to the top. Thus, the altitude might not be inside the parallelogram.

Recall that from any parallelogram we can cut off a triangular piece and move it to the other side to make it a rectangle. That is why the formula for the area of a parallelogram is so similar to the formula for the area of a rectangle.

Since any triangle is half of its corresponding parallelogram, the area of a triangle is half the area of that parallelogram:
$$A = \frac{bh}{2}$$
where b is the base and h is the altitude of the triangle.
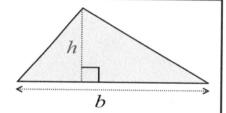

The **altitude** of a triangle is a line from one vertex to the opposite side that is perpendicular to that side. It can:

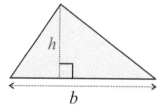   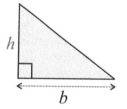   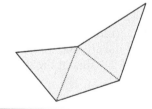

(1) fall inside the triangle;   (2) fall outside the triangle;   (3) be one of the sides of a right triangle.

To calculate the area of a polygon:

(1) Divide it into rectangles, triangles, and other simple shapes.
(2) Calculate the area of each part separately.
(3) Add the area of each of the parts.

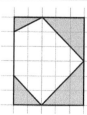
Sometimes we can use another strategy to find the area of a polygon.

(1) Draw a rectangle around the polygon.
(2) Calculate the areas of the triangles and quadrilaterals that are outside the polygon but inside the rectangle.
(3) Subtract those areas from the area of the entire rectangle.

You may use a calculator for all the problems in this lesson.

1. Find the area of the arrow figure, if each square in the grid measures 5 mm by 5 mm.

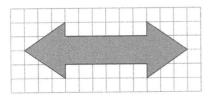

2. This is a pattern of light-colored rhombi with dark pink borders.

   The figure below shows the basic unit, or cell, of the pattern.
   The borders of the rhombus actually consist of parallelograms.

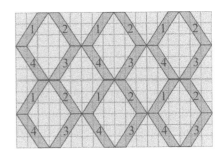

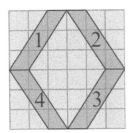

   a. Find the area in square units of one of the numbered parallelograms.

   b. Find the total area, in square units, of the lightly-colored parts of the cell.

   c. What percentage of the entire pattern do the rhombi cover?

   What percentage do the dark pink borders cover?

3. The sides of a triangle measure 4.7 cm, 9.8 cm, and 6.2 cm. Its altitude is 2.4 cm. The triangle is shrunk proportionally so that the altitude of the triangle becomes 1.4 cm. Find the area of the smaller triangle.

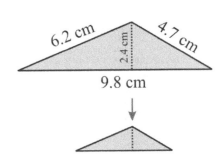

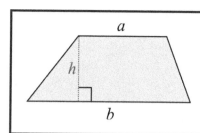

The **area of a trapezoid** is $A = \dfrac{(a+b)}{2} h$

where *a* and *b* are the lengths of the two parallel sides and *h* is the altitude. Essentially, we calculate the average of the lengths of the two parallel sides, and multiply that times the height.

4. Find the area of each trapezoid.

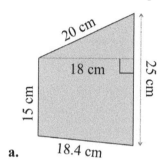

a.

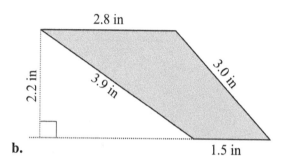

b.

5. The two parallel sides of a trapezoid measure 12 cm and 9 cm, and its altitude is 7 cm.

   a. Draw a trapezoid using this information, either on paper or in drawing software.

   b. Do the specified dimensions determine a single, unique trapezoid, or is it possible to draw more than one shape of trapezoid that satisfies those dimensions? Justify your answer.

6. A plot of land is in the shape of a trapezoid. The two parallel sides of the trapezoid measure 80 m and 60 m. The total area is 2,240 m². What is the altitude of the trapezoid?

Here's a reminder of how to draw an altitude to a triangle:

Line up the 90-degree mark on the protractor with the base of the triangle and the base of the protractor with the top vertex of the triangle. Draw the altitude along the base of the protractor.

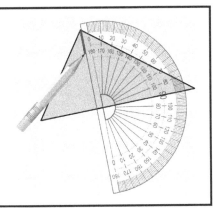

7. Find the area of this triangle.

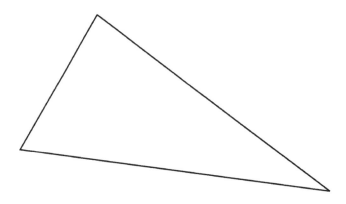

8. Find the area of this quadrilateral.

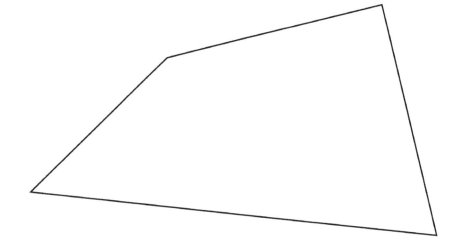

9. A swimming pool has the shape of the union of a square and a semicircle. The outer perimeter of the pool will be lined with decorative tiles that are 20 cm long. There will be 1 cm spaces between the tiles. The tiles are sold in packages of 20 that cost $24.90 per package.

Find how many packages of tiles must be purchased to line the perimeter of the pool and the total cost.

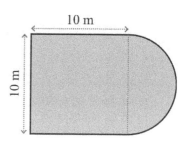

10. A garden has a circular pool of water surrounded by a hexagonal walking area.

    a. Calculate the area of the hexagon (including the pool). The diagram on the right shows the dimensions.

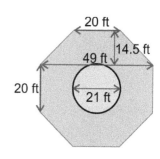

    b. Calculate the area of the hexagonal walking area not including the pool.

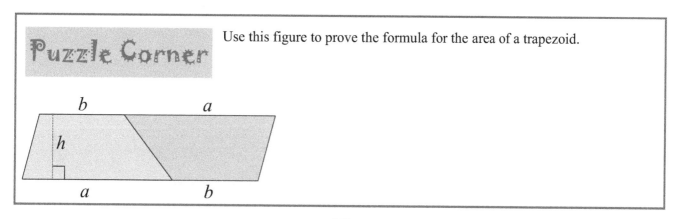

Puzzle Corner — Use this figure to prove the formula for the area of a trapezoid.

Surface Area

The **surface area** of a solid is the total area of all of its faces.

The **net** of a solid shows all the faces of the solid drawn in a plane, as if on a flat paper. It often makes calculating the surface area easier.

Example 1. The net of a cylinder consists of two circles for the bases and a rectangle for the lateral face (the face that "wraps around").

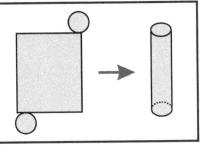

1. Sketch a net for each cylinder and calculate its surface area to the accuracy indicated.

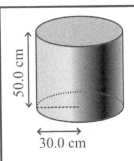

a. _____ cm²

Round your answer to the nearest hundred square centimeters (three significant digits).

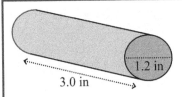

b. _____ in²

Round your answer to the nearest square inch (two significant digits).

154

2. Jeannie made a pyramid out of paper for a decoration. The bottom of the pyramid was a square with 5.0-cm sides, and the height of each triangular face of the pyramid was 4.0 cm.

   a. Sketch a net for Jeannie's pyramid.

   b. Calculate the surface area of the pyramid.

   c. (Optional). Make this pyramid from paper.

3. Find the surface area of this building (not including the bottom face).

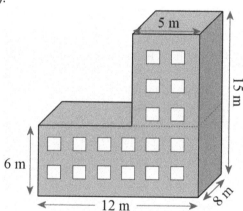

4. The surface area of a cube is 150 cm².
   Calculate the volume of the cube.

5. One cube has edges 1 unit long, and another has edges 2 units long.
   In other words, the lengths of their edges are in a ratio of 1:2.

   a. What is the ratio of their surface areas?

   b. What is the ratio of their volumes?

6. Three mirrors are set up in a form of an equilateral triangular prism.
   They will go inside a tube to make a kaleidoscope.

   Calculate the total surface area of the mirrors
   (not the end triangles).
   Round to the nearest ten square centimeters.

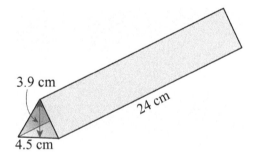

A half-circle can be folded into a cone. The folded part is called the **lateral face** of the cone: it is the face that "wraps around" the base. The base of a cone is a circle.

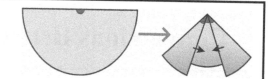

Example 2. Find the surface area of the cone with this net.

1. The bottom face is a circle with a radius of 2.25 cm. Its area is $\pi r^2 = \pi \cdot (2.25 \text{ cm})^2 \approx 15.9043 \text{ cm}^2$.

2. The lateral face is a half circle, with a radius of 4.5 cm. Its area is $(\frac{1}{2})\pi \cdot r^2 = 0.5 \cdot \pi \cdot (4.5 \text{ cm})^2 \approx 31.8086 \text{ cm}^2$.

Lastly we add the two: $15.9043 \text{ cm}^2 + 31.8086 \text{ cm}^2 = 47.7129 \text{ cm}^2 \approx 48 \text{ cm}^2$.

Notice that we carried several more decimals in the intermediate calculations than we kept for the final answer. For best accuracy, don't round your intermediate results—round only the final answer.

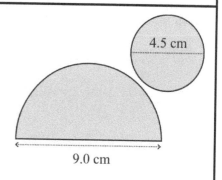

7. **a.** Draw a half-circle with a radius of 7.5 cm on blank paper. Cut it out and fold it into a cone.

   b. Calculate the surface area of your cone (without the base circle).

8. Find the surface area of this cone. Its lateral face is ¾ of a circle.

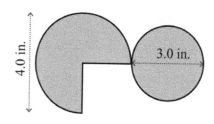

Puzzle Corner

Find the surface area of this figure.

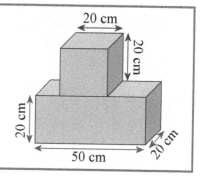

157

Conversions Between Customary Units of Area

The two tables below show the units of length and area in the customary system. Notice that **for each unit of length there is a corresponding unit of area**—but there is one unit of area that is different.

Units of Length

	Unit	Abbreviation
	mile	mi
1,760	yard	yd
3	foot	ft
12	inch	in

Units of Area

	Unit	Abbreviation	As a square/rectangle
	square mile	sq mi or mi^2	1 mi × 1 mi
640	acre	acre	660 ft × 66 ft
4,840	square yard	sq yd or yd^2	1 yd × 1 yd
9	square foot	sq ft or ft^2	1 ft × 1 ft
144	square inch	sq in or in^2	1 in × 1 in

I hope you noticed that *acres* are different—they don't have a corresponding unit of length. Instead, an **acre** is simply **4,840 square yards** or **43,560 square feet.** In the U.S. it's common to measure land by the acre.

Why is an acre such an odd number of square feet? The acre seems to have started as an approximation to the amount of land an ox could plow in one day. It was later defined to be a piece of land 660 ft long and 66 ft wide (660′ × 66′ = 43,560 sq ft). Today, an acre is a measure of area and does not have to be any particular shape.

Compared to an American football field, one acre would occupy the full width of the field (53 1/3 yards) and a bit over 90 of the 100 yards of its length.

1. Refer to the diagram of the square at the right to fill in this equation (It's really a conversion factor):

    1 square foot = _____ square inches.

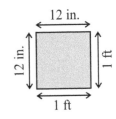

2. Sketch a square with sides 3 feet long. Mark its sides in inches, too.

    The area in square feet is _____ ft × _____ ft = _____ ft^2.

    The area in square inches is _____ in × _____ in = _____ in^2.

3. A rectangle measures 10 inches by 39 inches.

    a. Calculate its area in square inches.

    b. Convert the area into square feet, to the nearest tenth of a square foot, using the conversion factor from exercise 1.

Example 1. The sides of a rectangular piece of land measure 0.5 mi and 0.6 mi. How many acres is it?

First we calculate the area in square miles. It is 0.5 mi · 0.6 mi = 0.3 square miles. Then we use the conversion factor 1 square mile = 640 acres. There are two ways to think about this conversion:

(1) Square miles are bigger units than acres, so there are fewer of them than of acres. Therefore, we multiply the area by the number 640 and get 0.3 · 640 = 192 acres.

(2) We multiply the area 0.3 sq mi by a conversion ratio — either 1 sq mi/640 acres or 640 acres/1 sq mi. Since we want the square miles to cancel out, we choose 640 acres/1 sq mi to get:

0.3 sq mi · 640 acres/sq mi = 192 acres

4. Find the area in the given units. Then convert it to the other area unit. The conversion factors are found in the tables at the beginning of the lesson.

a. A rectangle with 4.0 ft and 2.5 ft sides

A = _____ sq ft

A = _____ sq ft · $\dfrac{144 \text{ sq in}}{1 \text{ sq ft}}$ = _____ sq in.

b. A rectangle with 20 ft and 15 ft sides

A = _____ sq ft

A = _____ sq ft · $\dfrac{\text{sq yd}}{\text{sq ft}}$ = _____ sq yd. (*to the nearest square yard*)

c. A rectangle with 0.4 mi and 0.9 mi sides

A = _____ sq mi

A = _____ sq mi · _____ = _____ acres. (*to the nearest ten acres*)

d. A square with 750 ft sides

A = _____ sq ft

A = _____ sq ft · _____ = _____ acres. (*to tenth of an acre*)

5. Find the area of this plot of land in acres.

320 ft
520 ft
340 ft
370 ft

159

6. Gardener Tom designed this geometric flower arrangement where the outer triangular areas have geraniums and the inner parallelogram has zinnias.

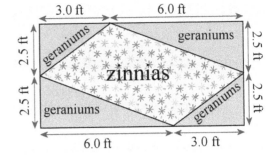

   a. What is the area, in square feet, reserved for geraniums?

   b. And for zinnias?

   c. Give the above two areas in square yards.

   d. What percentage of the entire flower arrangement do zinnias occupy?

7. Farmer Green wanted to plant a 560 ft by 800 ft plot with oats. He was advised to use 70 lb of nitrogen fertilizer per acre. How many pounds of fertilizer does he need?

8. Angie wants to buy some beneficial nematodes (a kind of worm) to get rid of ticks, fleas, termites, and other unwanted bugs from her yard. The catalog says that a package of 10 million nematodes treats 3,200 square feet of yard. The diagram at the right is a plan for Angie's yard.

   a. Find the area of the yard (not including the house) in square feet.

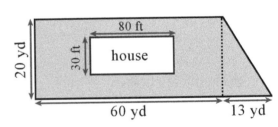

   b. Find how many packages of nematodes Angie needs.

Conversions Between Metric Units of Area

The two tables below show the units of length and the units of area in the metric system. Notice that **for each unit of length there is a corresponding unit of area**.

Units of length

	Unit	Abbrev.
10 ⤵	kilometer	km
10 ⤵	hectometer	hm
10 ⤵	decameter	dcm
10 ⤵	**meter**	**m**
10 ⤵	decimeter	dm
10 ⤵	centimeter	cm
	millimeter	mm

Units of area

	Area unit	abbrev.	As a square
100 ⤵	square kilometer	km^2	1 km × 1 km
100 ⤵	hectare	ha	100 m × 100 m
100 ⤵	are	a	10 m × 10 m
100 ⤵	**square meter**	**m^2**	**1 m × 1 m**
100 ⤵	square decimeter	dm^2	1 dm × 1 dm
100 ⤵	square centimeter	cm^2	1 cm × 1 cm
	square millimeter	mm^2	1 mm × 1 mm

Notice that the conversion factor between each two neighboring units of *length* is always 10, but the conversion factor between each two neighboring units of *area* is 100 (10 times 10).

The *hectare* is often used in measuring land. It is 10,000 m^2 which means that a square with 100-m sides is one hectare. However, keep in mind that a hectare is a unit for measuring area, not a particular shape. The total area of an American football field (including the end zones) is a little over 1/2 hectare.

1. The image shows 1 square centimeter.

    The area in square centimeters is _____ cm × _____ cm = _____ cm^2.

    The area in square *millimeters* is _____ mm × _____ mm = _____ mm^2.

2. Sketch a square with 1-meter sides (it doesn't have to be to scale).

    The area in square meters is 1 m × 1 m = 1 m^2.

    The area in square *decimeters* is _____ dm × _____ dm = _____ dm^2.

    The area in square *centimeters* is _____ cm × _____ cm = _____ cm^2.

    The area in square *millimeters* is _____ mm × _____ mm = _____ mm^2.

    So, 1 m^2 = _____ dm^2 = _____ cm^2 = _____ mm^2.

3. Convert 9 m^2 into square centimeters.

4. Find the area of this rectangle both in square centimeters and in square millimeters.

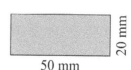

Example 2. A rectangular field measures 125 m by 670 m. What is its area in hectares?

Since the length and width are given in meters, we first find the area in square meters: 125 m · 670 m = 83,750 m². Then we convert that area into hectares. Here are two ways to make that conversion:

(1) The conversion factor between hectares and square meters is 1 ha = 10,000 m². We simply multiply the answer 83,750 m² either by the ratio 1 ha/10,000 m² or by the ratio 10,000 m²/1 ha. But which one?

We need a conversion factor to cancel the current unit of square meters and leave hectares in the answer. So hectares should be in the numerator and square meters in the denominator. Thus we multiply by the ratio 1 ha/10,000 m²:

$$83{,}750 \text{ m}^2 \cdot 1 \text{ ha}/10{,}000 \text{ m}^2 = 8.375 \text{ ha}$$

This of course means that we actually *divide* 83,750 by 10,000.

(2) Another way is to use estimation. Since the original rectangle measures *about* 100 m × 700 m, which would be exactly 7 hectares, the answer in hectares has to be near 7 ha. Therefore, 83,750 m² cannot equal 8,375 ha, nor 837.5 ha, nor 83.75 ha, nor 0.8375 ha — but must equal 8.375 hectares.

5. A rectangular field measures 500 m × 700 m. (Sketch it to help you.)
   Find its area in square meters and in hectares.

6. A village is spread out over a rectangular area with 0.2 km and 1.2 km sides.
   Calculate its area in square kilometers and in hectares.

7. The Louvre in Paris is the world's largest museum and houses one of the most impressive art collections in history. The exhibition floor space of the Louvre is 60,600 square meters. Convert this area to hectares.

Photo by Alvesgaspar

8. Jerry's favorite lake to fish in is roughly a rectangle that measures 1.5 cm by 0.8 cm on a map with a scale of 1:20,000. What is its approximate area in reality, to the nearest thousand square meters? To the nearest tenth of a hectare?

9. Find the area of the central ("field") part of this stadium in square meters and in hectares. Give your answers to four significant digits (to the nearest ten square meters and to the nearest thousandth of a hectare).

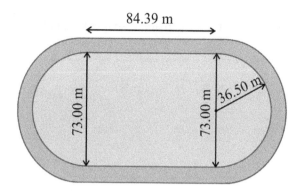

10. The town of Belfort in France has a land area of 17.1 km² and a population of about 51,000. Imagine that the land area of the whole town were divided equally among its citizens.

   a. How many square kilometers would each person get?

   b. How many hectares would each person get?

   c. How many square meters would each person get?

   d. The size of an average farm in France is about 50 ha. About how many times bigger is the average French farm than the average urban plot that each person in Belfort would get?

Slicing Three-Dimensional Shapes

In this lesson we examine the cross-sections that are formed when solids (three-dimensional shapes) are cut with a plane. But first, let's review some terminology.

- A **face** of a solid is a flat "side" with an area.
- An **edge** is the line segment where two faces meet.
- A **vertex** is a corner where three or more edges meet.

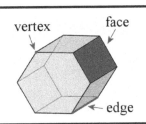

A **prism** has two identical polygons as the top and the bottom faces, each of which is called a **base**. Its other faces are parallelograms or rectangles.

Prisms are named after the base polygon. For example, a triangular prism has triangles as bases, and a quadrangular prism has quadrilaterals as bases.

Prisms are also classified as either **right** or **oblique** (slanted). In a right prism, the bases meet the other faces at right angles. Otherwise, the prism is oblique.

What we call a *box* in common language is called in mathematics a **right rectangular prism**: its bases are rectangles, and it is a right prism instead of oblique.

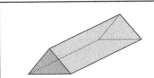

A right triangular prism

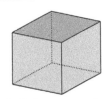

A right rectangular prism (a box)

An oblique triangular prism

An oblique pentagonal prism

A cylinder is like a prism, but its bases are not polygons. The figure commonly called a cylinder is more precisely a "right circular cylinder." Indeed, in mathematics, a cylinder can have a shape other than a circle as its base, and it can also be oblique instead of right.

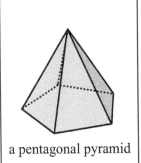

a pentagonal pyramid

A **pyramid** has a polygon as a base, and each vertex of that polygon is connected by an edge to a point called the top vertex or **apex**. The other faces of the pyramid are all triangles that meet at the top vertex. Pyramids are also named after the polygon at the base.

The **net** of a solid shows all the faces of the solid drawn on a flat surface.

The net of a pyramid is typically drawn with the base polygon in the middle and each triangular face connected to an edge of the base polygon.

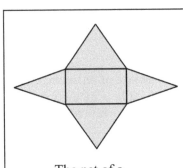

The net of a rectangular pyramid

1. Name the solids. Please note whether the solid is right or oblique.

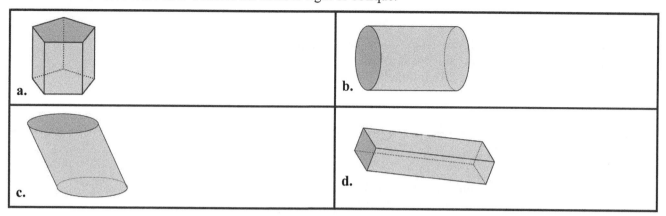

2. Name the solid that can be built from each net.

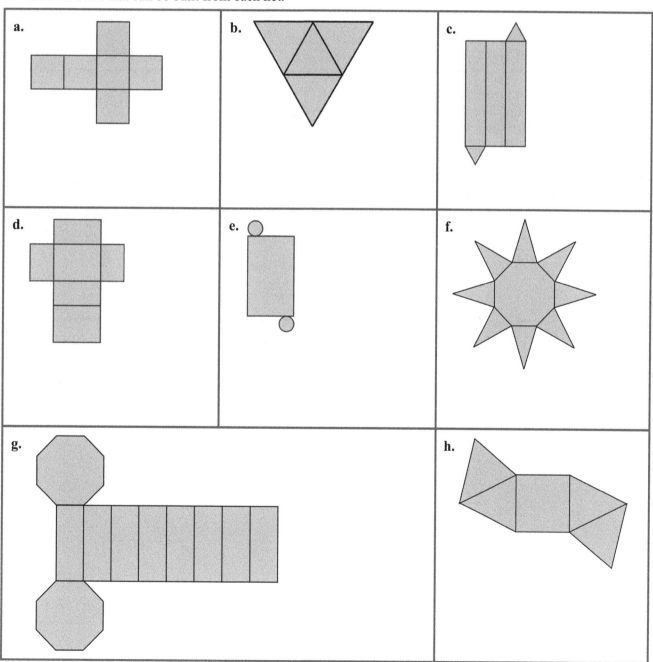

In geometry, a **plane** is a two-dimensional flat surface, with no thickness, that extends to infinity. It is like a sheet of paper that is infinitely thin and also infinitely long and wide.

Two planes can be parallel or intersect. Parallel planes are the same distance apart everywhere, so they never touch. If two planes are not parallel, then they will intersect in a line somewhere.

The first picture on the right shows a plane intersecting a cube vertically. The second picture shows the resulting **cross-section**, which is the two-dimensional shape formed by the intersection of the plane and the solid.

From our perspective, the cross-section may look like a parallelogram, but actually it is a square. You will verify that for yourself in an activity in this lesson.

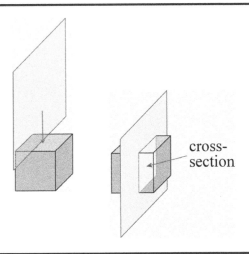

Example 1. A right cone is cut with a plane that is parallel to the base of the cone. The cross-section is a circle.

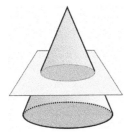

Example 2. A right rectangular prism is cut with a plane that is not parallel to the base of the prism. Even though the base is a square, the cross-section is a rectangle. Again, you may find it hard to visualize from the picture alone, but you will see it for yourself in one of the activities of this lesson.

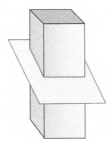

For the following three activities (questions 3, 4, and 5), you will need modeling clay, play-doh, dough, clay, or another similar substance that you can shape and cut with a knife. You will also need dental floss or a knife.

If you have Internet access, I highly recommend you also use the free online tool by the Shodor organization at

http://www.shodor.org/interactivate/activities/CrossSectionFlyer/

This tool shows cross-sections for various solids in an interactive manner. To get a cube or a rectangular prism, choose a "prism" and adjust the slider for "Lateral faces" until you get a cube.

3. **a.** Construct a cube from the clay. It doesn't have to be exact in its dimensions but should be as close to a cube as possible. In particular, it needs to have edges that are well-defined and sharp and as little rounded as possible.

   b. Slice the cube with dental floss or a knife and examine the cross-section (the cut surfaces) formed. Try several different kinds of cuts. Observe what kind of shapes the cross-sections are. Sketch each cross-sectional shape here:

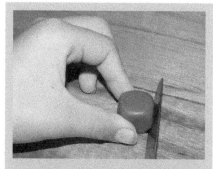

Use the flat side of a knife to help form the faces and edges of the cube.

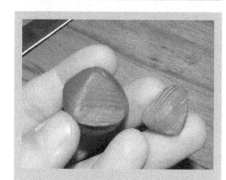

Can you find a cut so that the cross-section is a triangle?

   c. What shapes can you *not* get as a cross-section? Why not?

4. Repeat the activity with a right rectangular prism (box). Try several different kinds of cuts.

   a. Observe what kind of shapes the cross-sections are. Sketch each cross-sectional shape here:

What shape will the cross section from this cut be?

   b. Can you get all the same shapes from the cross-sections of this prism that you got from the cube? Explain.

5. Repeat the activity with a right rectangular pyramid. Try several different kinds of cuts. Observe what kind of shapes the cross-sections are. Sketch each cross-sectional shape here:

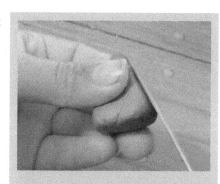

Again, to form the clay pyramid initially, it helps to press its faces with a knife.

6. Match each cut with its cross-section.

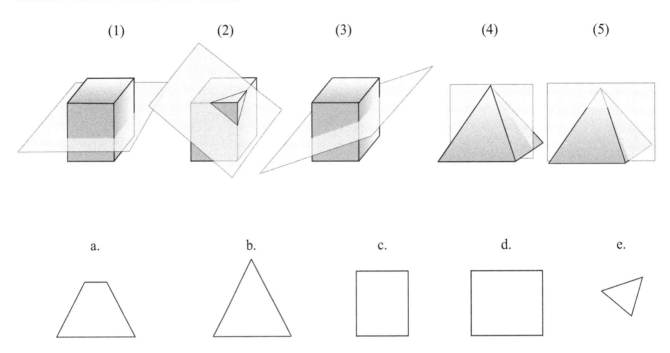

7. Describe the cross section formed by the intersection of the plane and the solid.

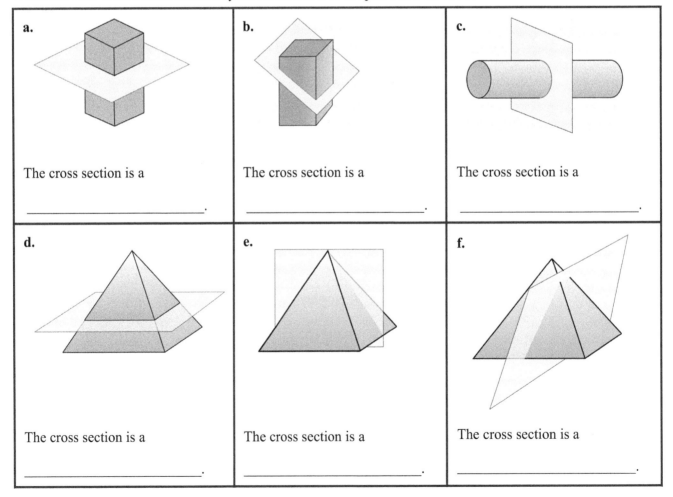

Volume of Prisms and Cylinders

The **volume** of both prisms and cylinders is calculated in the same way: we simply multiply the area of the base (A_b) times the height (h). The formula is: $V = A_b h$

Example 1. Calculate the volume of this right triangular prism.

This prism is lying "flat," so its height is 9 cm and its base (bottom figure) is a triangle. The area of the base triangle is:
A_b = (*base of triangle · height of triangle*/2) = 3 cm · 1.8 cm/2 = 2.7 cm².

Now we multiply the area of the base and the height to get the volume:
$V = A_b \cdot h$ = 2.7 cm² · 9 cm = 24.3 cm³ (cubic centimeters).

(Did you notice that the word "base" was used in two different ways?
1. The prism has a base, which is the shape of its <u>bottom face</u>.
2. The triangle has a base, which is its base <u>side</u>.)

Note: You may use a calculator for all problems in this lesson.

1. Calculate the volumes of these prisms and cylinders.

a.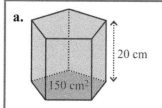

b. Round the volume to the nearest cubic foot.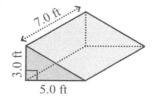

c.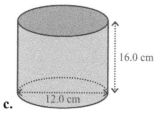

d.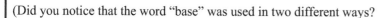

2. Calculate the volume of this little house to the nearest hundred cubic feet.

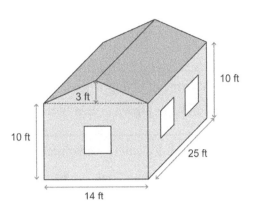

3. A large aquarium is in the shape of a trapezoidal prism. (The base of the prism is the trapezoid facing the viewer.)

   a. Calculate its volume. Round to three significant digits (a tenth of a cubic meter).

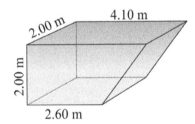

   b. If the aquarium is filled 95% full, how many liters of water does it hold?
      One cubic meter is 1000 liters.

> In metric units, it is easy to compare measures of liquid to units of volume. By definition, 1 milliliter equals 1 cubic centimeter, or **1 ml = 1 cm³**. Therefore, 1 liter = 1,000 cm³.

4. A rectangular juice carton measures 7.0 cm × 11.5 cm × 12.0 cm. Calculate its volume to the nearest ten milliliters. Give the volume in liters, too.

5. Find some container around your house. Measure the dimensions you need to calculate its volume in milliliters. If the container is empty, you can check your result by filling it with water and measuring the water in a measuring cup.

6. A cylindrical water pipe has an inside diameter of 2.50 cm. How much water can the pipe hold per meter of length? Give your answer in milliliters.

> Each edge of this cube measures 1 ft, which means its volume is **1 cubic foot**.
> Using the cube, it's very simple to figure out how many cubic inches are in one cubic foot: In cubic inches, its volume is 12 in · 12 in · 12 in = 1,728 in³.
> So, 1 ft³ = 1,728 in³.

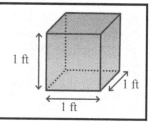

7. **a.** Use similar reasoning to find how many cubic inches there are in one cubic yard.

   b. Use similar reasoning to find how many cubic centimeters there are in a cubic meter.

8. A regular five-gallon bucket is approximately a circular cylinder with a height of 14.5 inches and a bottom (inner) diameter of 10.625 inches. (In reality, the top diameter is a little larger than the bottom diameter.)

   a. Calculate its volume in cubic inches.

   | 1 ft³ = 1,728 in³ |
   | 1 gallon = 231 in³ |

   b. Convert the volume to gallons. Is it exactly 5 gallons, less, or more?

   c. Convert the volume into cubic feet.

9. The inside dimensions of a chest freezer are
32 in (length) × 14 in (width) × 26 in (depth).
However, those dimensions don't give the true usable
volume because there is a box-shaped space inside which
contains the compressor. The height of that space is 8 in,
and its length is 12 in.

   a. Calculate the usable volume of the chest freezer
   in cubic inches.

   b. Convert that volume into cubic feet.

10. A swimming pool is in the shape of a rectangular prism.
It is 12.5 m long, 6.0 m wide, and 2.0 m deep. It is 7/8 full of water.

    $1 \text{ m}^3 = 1{,}000 \text{ L}$

    a. Calculate the volume of *water* (not of the pool) in cubic meters and in liters.

    b. It rains 15 mm. How many liters of water does that add to the pool?

Puzzle Corner

You make a paper cylinder using an 8.5 in by 11 in sheet of paper so that the cylinder stands 11 in tall. For taping, you overlap the edges of the paper by 1/2 inch. How big are the top and bottom circles? Calculate their diameter.

(Alternatively, you could use A4 size paper, which measures 210 mm × 297 mm, and overlap the edges by 1 cm.)

Chapter 8 Mixed Review

1. Give a real-life situation for the sum $(-30) + (-12)$.

2. Solve $53 + (-91) + 21 + (-3) + (-55)$.

3. On average, Scott makes a basket nine times out of twelve shots when he is practicing. How many baskets can he expect to make when he tries 200 shots?

   Use equivalent rates.

   $$\frac{9 \text{ baskets}}{12 \text{ shots}} = \frac{\boxed{} \text{ baskets}}{\boxed{} \text{ shots}} = \frac{\boxed{} \text{ baskets}}{200 \text{ shots}}$$

4. The two parallelograms are similar with a similarity ratio of 2:7. Find the length of the unknown side.

   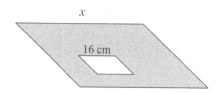

5. A book costing $15 is discounted by 15%. What is the new price?

6. At 20% off, a tire costs $100. What was its price before the discount?

7. A certain county acquired a total revenue of $135,000 in year 2013. Of it, 41.0% came from property tax and the rest from other sources. In 2014, the total revenue fell by $3,500 and the revenue from property tax by $2,100. What percentage of the county's revenue came from property tax in 2014?

8. Simplify the expressions.

| a. $2 + w + 11 + w + 2w$ | b. $2 \cdot w \cdot 11 \cdot w \cdot 2w$ | c. $c \cdot c \cdot 3 \cdot d \cdot d \cdot d$ |

9. Car 1 gets a gas mileage of 20 miles per gallon and car 2 gets 24 miles per gallon.

   a. Write two equations—one for each car—relating the distance (d) driven to the amount of gasoline used. Use g to represent the amount of gasoline.

   Car 1: $d =$ _____   Car 2: $d =$ _____

   b. Plot both equations in the same coordinate grid.

   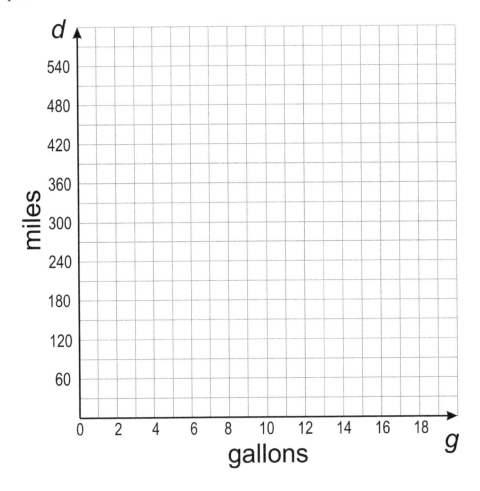

   c. State the slopes of the two lines.

   d. Plot a point on each line that corresponds to the distance 120 miles.

   e. Plot a point on each line that corresponds to the unit rate.

   f. What does the point (0, 0) mean in this situation?

10. To the nearest tenth of a percent, calculate by how many percent these quantities changed.

a. The postage for a letter increased from $0.78 to $0.81.	b. Jamie sold 445 newspapers last week. This week he sold 487.

11. A rectangle with sides of 2 1/4 in. and 3 in. is enlarged by a scale factor of 3.5. Find the area of the resulting rectangle to the tenth of a square inch.

12. How many 3/4-foot long pieces can you cut out of a roll of string that is 8 3/8 feet long? Write a *quotient* to solve the problem.

  ———— =

13. One of the expressions below gives an area of a rectangle and the other its perimeter.

   $14a + 8b$ $7a \cdot 4b$

   How long is each side of the rectangle?

14. Solve.

a. $1\dfrac{1}{3} - y = \dfrac{5}{8}$	b. $z + \dfrac{2}{3} = 1\dfrac{9}{10}$

Chapter 8 Review

1. The letters from *u* to *z* label the angles in the figure.

    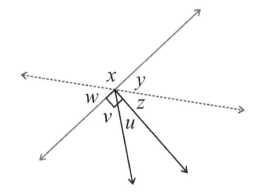

    a. Which two angles are complementary?

    b. Which two angles are supplementary?

    c. Which two angles are vertical angles?

    d. If $w = 51°$ and $v = 59°$, then what are the measures of the rest of the angles?

    $u = $ _____ ° $x = $ _____ ° $y = $ _____ ° $z = $ _____ °

2. In the figure at the right, lines *l* and *m* are parallel.

    a. Write an equation for the measure of the unknown angle *x*.

    b. Solve your equation.

    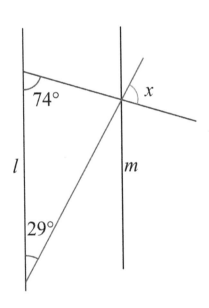

3. The top angle of an isosceles triangle measures 26°.

   a. What do the other two angles measure?

   b. Draw the triangle. Make the base 4 inches long.

4. Draw, using a compass and straightedge only, an isosceles triangle with two sides the length of this line segment:

5. Draw a triangle with sides 2 in, 2 3/4 in, and 3 3/8 in long.

6. Two sides of a parallelogram measure 10.2 cm and 5.0 cm. There is a 45° angle between them. Do these conditions define a unique parallelogram? If so, draw it. If not, draw several non-congruent (different-shaped) parallelograms that all fit the conditions.

7. This rectangle is a plan for Henry's room drawn at a scale of 1:50. Draw a copy of it at the scale 1:60.

Scale 1:50

8. Draw an altitude into this triangle and then find its area to the nearest square centimeter.

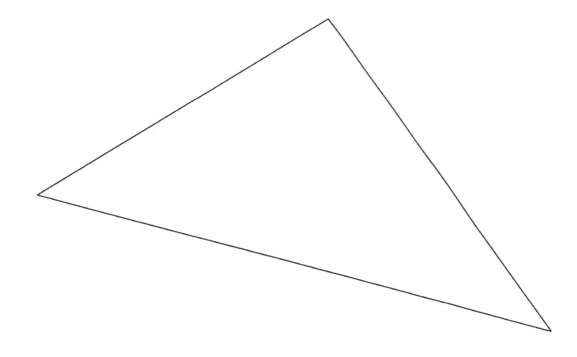

9. **a.** Find the radius of a circle with a circumference of 20.6 cm.

   b. Calculate, to the nearest ten square inches, the area of a circle with a diameter of 8 ft 2 in.

10. Explain in a few words what the pictures at the right are about.

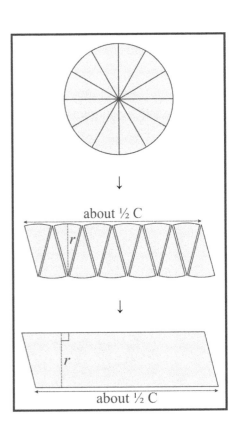

11. The roof of a canopy is in the form of a pyramid. Calculate the total surface area of the roof.

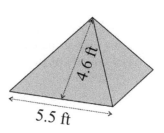

12. Calculate the volume of this box to the nearest hundred cubic inches.

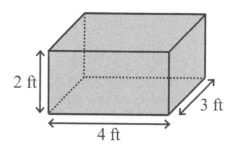

13. **a.** Find the area of this trapezoid in square feet.

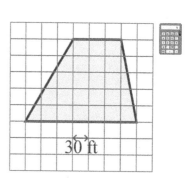

   b. Find the area in square yards, too.

14. A glass jar is approximately in the shape of a circular cylinder. Jamie measured its volume using water and a measuring cup, and he found that it held about 530 ml of water. Jamie also measured its diameter as 9 cm and its height as 12 cm.

   a. How many cubic centimeters is 530 ml?

   b. One of Jamie's measurements—either the diameter or height—is in error. How can you tell that?

15. The packaging for a granola bar is in the form of a trapezoidal prism.

   a. Find its volume in cubic millimeters.

   b. Find its volume in cubic centimeters.

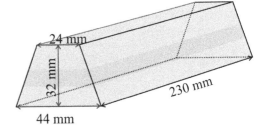

16. What shape is the cross-section?

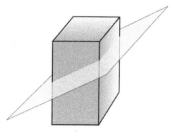

17. Explain or draw a sketch of how to cut through this pyramid with a plane so that the cross-section is a triangle.

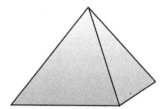

Chapter 9: The Pythagorean Theorem
Introduction

This is a relatively short chapter focusing on the Pythagorean Theorem and its applications. The Pythagorean Theorem is actually not part of the Common Core Standards for seventh grade. The Common Core places it in eighth grade. However, I have included it in this curriculum because it is a traditional topic in pre-algebra. That way, Math Mammoth Grade 7 works as a full pre-algebra curriculum while fully meeting (and exceeding) the Common Core Standards for grade 7. If you are following the Common Core Standards strictly, you can safely omit this entire chapter, because your student(s) will encounter these topics in eighth grade.

First, students need to become familiar with square roots, so they can solve the equations that result from applying the Pythagorean Theorem. The first lesson of the chapter introduces taking a square root as the opposite operation to squaring a number. The lesson includes both applying a guess-and-check method and using a calculator to find the square root of a number.

Next, students learn how to solve simple equations that include taking a square root. This makes them fully ready to study the Pythagorean Theorem and apply it.

The Pythagorean Theorem is introduced in the lesson by that name. Students learn to verify that a triangle is a right triangle by checking if it fulfills the Pythagorean Theorem. They apply their knowledge about square roots and solving equations to solve for an unknown side in a right triangle when two of the sides are given.

Next, students solve a variety of geometric and real-life problems that require the Pythagorean Theorem. This theorem is extremely important in many practical situations. Students should show their work for these word problems to include the equation that results from applying the Pythagorean Theorem to the problem and its solution.

There are literally hundreds of proofs for the Pythagorean Theorem. In this chapter, we present one easy proof based on geometry (not algebra). As an exercise, students are asked to supply the steps of reasoning to another geometric proof of the theorem, and for those interested the lesson also provides an Internet link that has even more proofs of this theorem.

You can find matching videos for topics in this chapter at http://www.mathmammoth.com/videos/ (grade 7).

The Lessons in Chapter 9

	page	span
Square Roots	189	*4 pages*
Equations That Involve Taking a Square Root	193	*5 pages*
The Pythagorean Theorem	198	*5 pages*
The Pythagorean Theorem: Applications	203	*7 pages*
A Proof of the Pythagorean Theorem	210	*1 page*
Chapter 9 Mixed Review	211	*3 pages*
Chapter 9 Review	214	*3 pages*

Helpful Resources on the Internet

The Pythagorean Theorem - Video Lessons by Maria
A set of free videos that teach the topics in this chapter - by the author herself.
http://www.mathmammoth.com/videos/prealgebra/pre-algebra-videos.php#pythagorean

SQUARE ROOTS

Squares and Square Roots
A fun lesson about squares and square roots with lots of visuals and little tips. It is followed by 10 interactive multiple-choice questions.
http://www.mathsisfun.com/square-root.html

The Roots of Life
Practice finding square roots of perfect squares and help the roots of a tree grow.
http://www.hoodamath.com/games/therootsoflife.html

Square Root Game
Match square roots of perfect squares with the answers. Includes several levels.
http://www.math-play.com/square-root-game.html

Approximating Square Roots
Practice finding the approximate value of square roots by thinking about perfect squares.
https://www.khanacademy.org/math/pre-algebra/pre-algebra-exponents-radicals/pre-algebra-square-roots/e/square_roots_2

Equations with Square Roots & Cube Roots
Test your knowledge of square roots and cube roots with this interactive online quiz.
https://www.khanacademy.org/math/cc-eighth-grade-math/cc-8th-numbers-operations/cc-8th-roots/e/equations-w-square-and-cube-roots

Pyramid Math
Choose "SQRT" to find square roots of perfect squares. Drag the correct answer to the jar on the left. This game is pretty easy.
http://www.mathnook.com/math/pyramidmath.html

Rags to Riches Square Root Practice
Answer multiple-choice questions that increase in difficulty. The questions include finding a square root of perfect squares, determining the two nearest whole numbers to a given square root, and finding square roots of numbers that aren't perfect squares to one decimal digit.
http://www.quia.com/rr/382994.html

THE PYTHAGOREAN THEOREM

Pythagorean Theorem - Braining Camp
This learning module includes a lesson, an interactive manipulative, multiple-choice questions, real-life problems, and interactive open-response questions.
https://www.brainingcamp.com/content/pythagorean-theorem/

Pythagoras' Theorem from Maths Is Fun
A very clear lesson about the Pythagorean Theorem and how to use it, followed by 10 interactive practice questions.
http://www.mathsisfun.com/pythagoras.html

Pythagorean Triplets
Move the two orange points in this activity to find Pythagorean Triplets, sets of three whole numbers that fulfill the Pythagorean Theorem.
http://www.interactive-maths.com/pythagorean-triples-ggb.html

The Pythagorean Theorem Quiz
A 10-question quiz that asks for the length of the third side of a right triangle when the two sides are given.
http://www.thatquiz.org/tq-A/?-j10-la-p1ug

Interactivate: Pythagorean Theorem
Interactive practice problems for calculating the third side of a right triangle when two sides are given.
http://www.shodor.org/interactivate/activities/PythagoreanExplorer/

Exploring the Pythagorean Theorem
This multimedia mathematics resource shows how the Pythagorean Theorem is an important math concept used in the structural design of buildings. Using an interactive component, students construct right triangles of various sizes to explore calculations of the Pythagorean Theorem.
http://www.learnalberta.ca/content/mejhm/index.html?l=0&ID1=AB.MATH.JR.SHAP&ID2=AB.MATH.JR.SHAP.PYTH

Pythagorean Theorem Challenge
Review the Pythagorean Theorem in this interactive self-check quiz.
https://www.khanacademy.org/math/basic-geo/basic-geometry-pythagorean-theorem/pythagorean-theorem-app/e/pythagorean-theorem-word-problems

Pythagorean Theorem Test
Test your knowledge of the Pythagorean Theorem in this interactive online quiz.
http://www.mathportal.org/math-tests/trigonometry-tests/tests-in-right-triangle-trigonometry.php?testNo=1&testName=Pythagorean-Theorem-Test

Pythagoras in 3D
A challenge problem: can you find the longest dimension of a box?
http://www.interactive-maths.com/pythagoras-in-3d-ggb.html

PROOF

Proving the Pythagorean Theorem
See if you can figure out two more proofs of the Pythagorean theorem. Only the pictures are given to you. Tips and Solutions are available.
http://www.learner.org/courses/learningmath/geometry/session6/part_b/more.html

Annotated Animated Proof of the Pythagorean Theorem
Watch the animation to learn a proof of the Pythagorean Theorem.
http://www.davis-inc.com/pythagor/proof2.html

Many Proofs of the Pythagorean Theorem
A list of animated proofs.
http://www.takayaiwamoto.com/Pythagorean_Theorem/Pythagorean_Theorem.html

Pythagorean Theorem and its many proofs
A collection of 111 approaches to prove this theorem.
http://cut-the-knot.com/pythagoras/

Square Roots

The **square** of a number is that number multiplied by itself:

$$\text{six squared} = 6^2 = 6 \cdot 6 = 36$$

Simply put, the square of 6 tells you the area of a square with sides 6 units long.

Taking a **square root** is the opposite operation to squaring. For example, the square root of 36 is 6. This operation goes the opposite way: if you know the area of a square, then you can find the length of its side.

We use the "$\sqrt{}$" symbol (called the "radical") to signify "square root."
For example, $\sqrt{25} = 5$ because $5^2 = 25$.

Here is a way to help you remember what a square root is. In the picture on the right, the area of a square is written inside the square and the length of the side is written to the side:

Now, imagine the square is a square root symbol that "houses" the number for the area:

To find a square root of a number, think of a square with that area, and find the length of the side of that square.

$$\sqrt{49} = 7$$

1. Find the square roots.

a. $\sqrt{100}$	b. $\sqrt{64}$	c. $\sqrt{4}$	d. $\sqrt{0}$
e. $\sqrt{81}$	f. $\sqrt{144}$	g. $\sqrt{1}$	h. $\sqrt{10{,}000}$

2. It is especially easy to find square roots of numbers that are **perfect squares**: numbers we get by squaring whole numbers. For example, 49 is a perfect square because it is 7^2.
Fill in the list of perfect squares from 1^2 to 20^2 at the right:

3. Now find these square roots. You can use the table at the right or guess and check.

a. $\sqrt{169}$ b. $\sqrt{900}$

c. $\sqrt{225}$ d. $\sqrt{121}$

e. $\sqrt{441}$ f. $\sqrt{8{,}100}$

Perfect squares

1	____
4	____
9	169
16	196
25	____
36	256
49	289
____	324
____	361
____	400

4. Solve and find a shortcut for simplifying expressions of the form $\sqrt{a^2}$.

a. $\sqrt{6 \cdot 6}$	b. $\sqrt{7^2}$	c. $\sqrt{57^2}$	d. $\sqrt{0.29^2}$

Fill in the shortcuts: Since squaring and square root are opposite operations,

$(\sqrt{a})^2 =$ ____ and $\sqrt{a^2} =$ ____ for any positive number a.

On the previous page you saw a list of numbers that were perfect squares (1, 4, 9, 16, 25, ...). The square roots of those numbers are whole numbers. However, most numbers, such as 2, 5, and 17, are not perfect squares, and their square roots are not so "pretty." In fact, their square roots are **irrational numbers**, which means they are unending decimals without any repeating patterns in the digits. We can use squaring with the guess-and-check technique to approximate their values.

Example 1. Find the value of $\sqrt{19}$ to two decimal digits.

First we find two consecutive perfect squares so that 19 is between them: $16 < 19 < 25$. From that fact we know that $4 < \sqrt{19} < 5$. Also, since 19 is closer to 16 than to 25, we would expect $\sqrt{19}$ to be closer to 4 than to 5. So let's choose $\sqrt{19} = 4.3$ or 4.4 as our initial guesses, square the guesses, and check how close to 19 we get:

$$4.3^2 = 18.49$$

$$4.4^2 = 19.36$$

We can see that $\sqrt{19}$ is between 4.3 and 4.4, and that it is closer to 4.4 than it is to 4.3 (because 19.36 is closer to 19 than 18.49 is). Let's try 4.36 next.

$4.36^2 = 19.0096$ This is *very* close to 19! It is just a bit big, so let's check the next smaller one, 4.35^2:

$$4.35^2 = 18.9225$$

Now we know that $\sqrt{19}$ is between 4.35 and 4.36 and closer to 4.36 than it is to 4.35 (because 19.0096 is much closer to 19 than 18.9225 is). This means that to two decimal digits, $\sqrt{19} = 4.36$.

5. Use only multiplication (squaring) to guess and check the values of the following square roots to two decimal digits. You may use a calculator, but not the calculator's "square root" function.

   a. $\sqrt{7}$

   b. $\sqrt{51}$

   c. $\sqrt{99}$

Calculators have a button with the radicand symbol "√" for calculating square roots. On some calculators, you first push the square root button, then the number of which you are taking the square root. On others, you first enter the number and then push the square root button. Find out which way your calculator works.

Remember, if a square root is not a whole number, it is an irrational number, and irrational numbers are unending decimals without any pattern in the digits. This means the calculator will show you only a *part* of the decimal expansion of a square root—as many digits as fit onto its screen. For example, you might see:

$$\sqrt{14} = 3.7416573867739413855837487323165$$

6. Use a calculator to find these square roots. Round your answers to four decimal digits.

a. $\sqrt{8}$	b. $\sqrt{12}$	c. $\sqrt{15.39}$
d. $\sqrt{5,493.2}$	e. $\sqrt{0.6}$	f. $\sqrt{0.01}$

The square root symbol acts as a grouping symbol: it is as if there were parentheses around the expression under the square root. In other words, $\sqrt{15 + 10}$ means $\sqrt{(15 + 10)}$.

Example 2. Simplify $\sqrt{5 \cdot (70 + 10)}$.

We simplify the expression under the square root first and take the square root last:

$$\sqrt{5 \cdot (70 + 10)} = \sqrt{5 \cdot 80} = \sqrt{400} = 20$$

7. Calculate.

a. $\sqrt{9 + 16}$	b. $\sqrt{11 \cdot 11}$	c. $\sqrt{2 \cdot (41 - 9)}$
d. $\sqrt{225 - 9^2}$	e. $\sqrt{10^2 - 8^2}$	f. $\sqrt{13^2 - 12^2}$

8. Find the value of these expressions to three decimal digits. Use a calculator. Note: if your calculator doesn't automatically follow the order of operations, you need to use parentheses when entering the expressions. Another option is to write the intermediate results down or load them into the calculator's memory.

a. $\sqrt{5.6^2 - 2.1^2}$	b. $\sqrt{45.7^2 + 38.12^2}$

There is something special about square roots and negative numbers. Try $\sqrt{-25}$ with a calculator. Surprising?

Can you imagine a square with a *negative* area? Using our previous illustration for area of a square, could you have $\boxed{-49}$?

Clearly, that is not possible. No matter how long or short the side of the square is, when you multiply it by itself, you always get a *positive* number! That is why we cannot take a square root of a negative number.

(More specifically, the square root of a negative number is not a real number. In high school math courses you will learn that mathematicians have found a way to get around this limitation by using *imaginary numbers*.)

Taking a square root of a negative number is not possible!

9. Simplify or state that the result is not a real number.

a. $\sqrt{300 + 600}$	b. $\sqrt{-49}$	c. $\sqrt{2-3}$
d. $\sqrt{25 - 6^2}$	e. $\sqrt{26^2 - 24^2}$	f. $\sqrt{35^2 - 39^2}$

10. **a.** What is the area of a square if its side measures $\sqrt{1,600}$ cm?

b. What is the area of a square if its side measures $\sqrt{37}$ in?

11. **a.** Sketch a square with an area of 18 square centimeters.

   b. What is its perimeter, to two decimal digits?

12. **a.** Sketch a square with a perimeter of 18 cm.

   b. What is its area, to two decimal digits?

Equations That Involve Taking a Square Root

Example 1. Solve $x^2 = 81$.

We can use mental math: one obvious solution is $x = 9$. However, there is also another solution! It is not only true that $9^2 = 81$, but $(-9)^2 = 81$ also, so $x = -9$ is a second solution to this equation.

Example 2. Solve $x^2 = 48$.

This time, we cannot solve the equation with mental math, but we will *take a square root of both sides of the equation*. This will undo the squaring, and isolate x, because taking a square root and squaring are opposite operations.

$x^2 = 48$

$x = \sqrt{48} \approx 6.93$

or $x = -\sqrt{48} \approx -6.93$

$\sqrt{}$ The radicand symbol signifies taking a square root of both sides of the equation.

Since taking a square root undoes the squaring, x is now left alone on the left side. Notice that there are two solutions: the square root of 48 and the negative square root of 48.

Notice that $-\sqrt{48}$ doesn't mean that we take a square root of a negative number. Instead, $-\sqrt{48}$ means we *first* take the square root of 48 (a positive number) and then take the opposite of that result.

1. Solve. Remember, there will be two solutions. When the solutions aren't integers, give them both as the square root of an integer and also as a decimal approximation rounded to two decimal digits.

a.	$x^2 = 25$	**b.**	$y^2 = 3{,}600$
c.	$x^2 = 500$	**d.**	$z^2 = 11$
e.	$w^2 = 287$	**f.**	$q^2 = 1{,}000{,}000$

Example 3. $x^2 + 78 = 129$ We want to isolate the term x^2, so we first subtract 78 from both sides.

$x^2 = 51$ Now we take a square root of both sides.

$x = \sqrt{51}$ or $x = -\sqrt{51}$ There are two solutions, as usual.

If this is strictly a math problem and does not involve quantities with units, the answer can be left in the square root form. Otherwise, you should find its decimal approximation.

Here are the checks. Usually, it is enough to check only the positive root ($x = \sqrt{51}$), as the check for the negative root ($x = -\sqrt{51}$) is practically identical.

$(\sqrt{51})^2 + 78 \stackrel{?}{=} 129$ $(-\sqrt{51})^2 + 78 \stackrel{?}{=} 129$

$51 + 78 \stackrel{?}{=} 129$ $51 + 78 \stackrel{?}{=} 129$

$129 = 129$ ✓ $129 = 129$ ✓

Example 4. $3x^2 = 40$ $\div 3$ Again, we want to isolate the term x^2, so we first divide both sides by 3.

$x^2 = 40/3$ $\sqrt{}$ This symbol signifies taking a square root of both sides of the equation.

$x = \sqrt{40/3}$ or $x = -\sqrt{40/3}$ There are two solutions, as usual.

$x \approx 3.651$ or $x \approx -3.651$ These are the decimal approximations.

Here is a check using the rounded positive root:

$3 \cdot 3.651^2 \stackrel{?}{=} 40$

$3 \cdot 13.329801 \stackrel{?}{=} 40$

$39.989403 \approx 40$ ✓

Here is a check using the exact positive root:

$3 \cdot (\sqrt{40/3})^2 \stackrel{?}{=} 40$

$3 \cdot 40/3 \stackrel{?}{=} 40$

$40 = 40$ ✓

2. Solve. Remember, there will be two solutions. Check your solutions.

a. $5x^2 = 125$

b. $y^2 + 100 = 1{,}000$

3. Solve. You can use a calculator. Give the final solution both in square root format and as a decimal approximation rounded to three decimals. For the next lesson (The Pythagorean Theorem) you need to be able to solve equations like these that involve square roots.

a. $\quad a^2 - 8 = 37$	b. $\quad 8.2b^2 = 319$
c. $\quad a^2 + 4.5 = 10.7$	d. $\quad 12b^2 = 36{,}000$

Example 5. $x^2 + 7^2 = 12^2$ First simplify.

$x^2 + 49 = 144$ Now it looks more familiar. Subtract 49.

$x^2 = 95$ Now we take a square root of both sides.

$x = \sqrt{95}$ or $x = -\sqrt{95}$ There are two solutions, as usual.

Check:
$(\sqrt{95})^2 + 7^2 \stackrel{?}{=} 12^2$
$95 + 49 \stackrel{?}{=} 144$
$144 = 144$ ✓

4. Solve. Round the answers to three decimals. Check your solutions. You can use a calculator.

a. $a^2 + 3^2 = 7^2$

b. $43^2 + x^2 = 51^2$

c. $s^2 = 2.1^2 + 5.4^2$

d. $21^2 + 29^2 = w^2$

5. Here are some more practice problems. Round the answers to three decimals.

a. $45 - x^2 = 20$	b. $112^2 + s^2 = 18{,}200$
c. $s^2 = 0.89^2 + 1.22^2$	d. $6{,}650 - y^2 = 70^2$

Puzzle Corner

Solve $x^2 - x = 0$.

The Pythagorean Theorem

You will now learn a very famous mathematical result, the Pythagorean Theorem, which has to do with the lengths of the sides in a right triangle. First, we need to study some terminology.

In a right triangle, the two sides that are perpendicular to each other are called **legs**. The third side, which is always the longest, is called the **hypotenuse**.

In the image on the right, the sides a and b are the legs, and c is the hypotenuse.

Note: We don't use the terms "leg" and "hypotenuse" to refer to the sides of an acute or obtuse triangle — this terminology is restricted to *right* triangles.

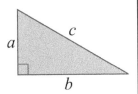

The Pythagorean Theorem states that **the sum of the squares of the legs equals the square of the hypotenuse.**

In symbols it looks much simpler:

$$a^2 + b^2 = c^2$$

The picture shows squares drawn on the legs and on the hypotenuse of a right triangle. Verify visually that the total area of the two yellow squares drawn on the legs looks about equal to the area of the blue square on the hypotenuse.

We will prove this theorem in another lesson.
For now, let's get familiar with it and learn how to use it.

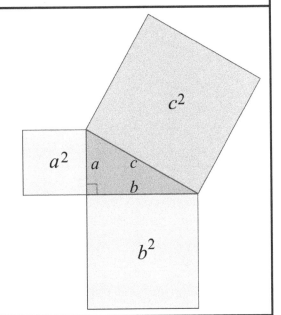

1. This is the famous 3-4-5 triangle: its sides measure 3, 4, and 5 units. It is a right triangle. Check that the Pythagorean Theorem holds for it by filling in the numbers below.

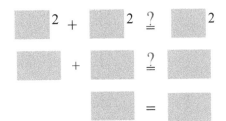

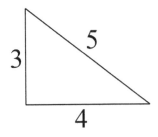

2. **a.** Check that the Pythagorean Theorem holds for a triangle with sides 6, 8, and 10 units long by filling in the numbers at the right.

   b. Use a compass and a ruler to draw a triangle with sides 6, 8, and 10 cm long. You can review the box, "A Triangle with Three Given Sides," on page 127. Measure its angles: did you get a right triangle?

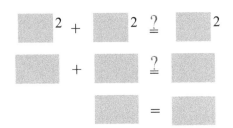

Example 1. This triangle is *not* a right triangle, so the Pythagorean Theorem does *not* hold:

$$2.55^2 + 3.31^2 \stackrel{?}{=} 3.58^2$$

$$6.5025 + 10.9561 \stackrel{?}{=} 12.8164$$

$$17.4586 > 12.8164$$

The sum of the areas of the squares drawn on the two shortest sides is more than the area of the square drawn on the longest side. As you can see, the triangle is acute.

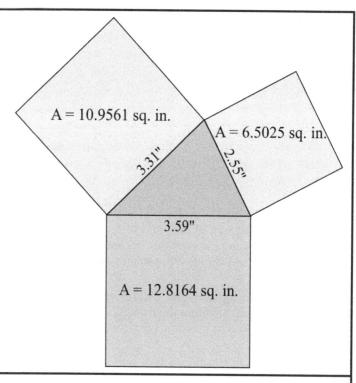

Example 2. Is a triangle with sides 4 cm, 5 cm, and 7 cm a right triangle?

We check if 4, 5, and 7 fulfill the Pythagorean Theorem (on the right). They don't. In fact, $4^2 + 5^2 < 7^2$ and the triangle is obtuse. (You can check that by drawing it.)

$$4^2 + 5^2 \stackrel{?}{=} 7^2$$

$$16 + 25 \stackrel{?}{=} 49$$

$$41 < 49$$

This triangle is obtuse.

3. For each set of lengths, determine whether they form a right triangle using the Pythagorean Theorem. Notice carefully which length is the hypotenuse.

   a. 6, 9, 13

   b. 12, 13, 5

4. **a.** Measure each side of this triangle to the nearest millimeter.

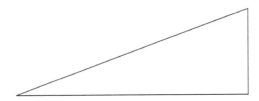

b. Verify that the sum of the areas of the squares on the legs is *very close* to the area of the square on the hypotenuse. I say "very close" because the process of measuring is always inexact, and therefore your calculations and results will probably not yield true equality, just something close.

5. For each set of lengths below, determine whether the lengths form an acute, right, or obtuse triangle—or *no* triangle. You can construct the triangles using a compass and a ruler and also use the Pythagorean theorem.

   a. 9, 6, 4

   b. 13, 11, 10

   c. 12, 14, 28

   d. 15, 20, 25

Example 3. The two legs of a right triangle measure 7.0 in. and 10.0 in. How long is the hypotenuse?

Let x be the length of the unknown side. We use the Pythagorean Theorem to solve for x:

$$7^2 + 10^2 = x^2$$
$$49 + 100 = x^2$$
$$x^2 = 149$$
$$x = \sqrt{149} \text{ or } x = -\sqrt{149}$$

In this case, we ignore the negative root as the length of a side cannot be negative!

$$x \approx 12.2 \text{ in.}$$

The hypotenuse measures about 12.2 in.

Example 4. Find the unknown leg of this right triangle.

This time we know the hypotenuse and one of the legs. The Pythagorean Theorem gives us:

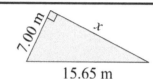

$$7.00^2 + x^2 = 15.65^2$$
$$49 + x^2 = 244.9225$$
$$x^2 = 195.9225$$

We keep all the decimals for the intermediate results.

$$x = \sqrt{195.9225} \text{ or } x = -\sqrt{195.9225}$$

Again, we ignore the negative root.

$$x \approx 14.0 \text{ m.}$$

The other leg measures about 14.0 m.

6. Solve for the unknown side of each right triangle to one decimal digit.

a.

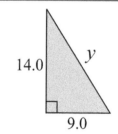

b.

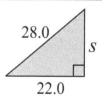

7. Find the length of the unknown side. Round your final answer to the same accuracy as the numbers in the problem.

a.

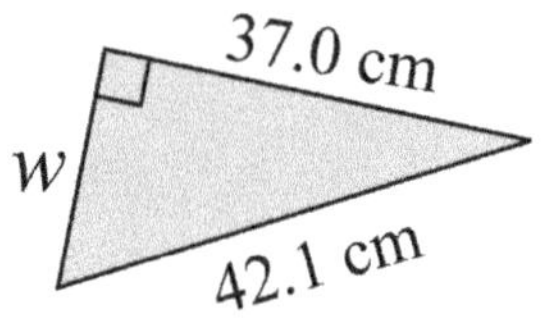
37.0 cm
w
42.1 cm

b.

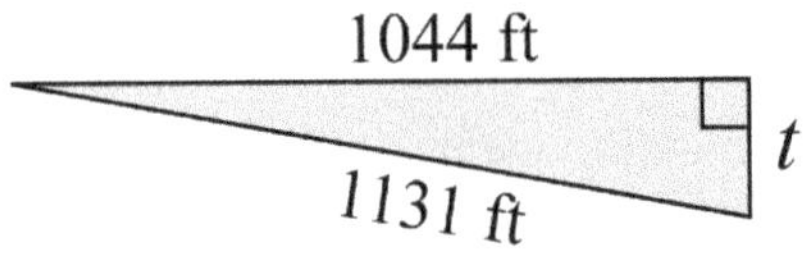
1044 ft
t
1131 ft

8. If the legs of a right triangle measure 12 ft 5 in and 7 ft 8 in, find the length of the hypotenuse to the nearest inch.

Puzzle Corner

A math teacher made the problem below for a test. Find what went wrong with it. Then fix the problem, so it can be used in the test, and solve it.

How long is the unknown side?

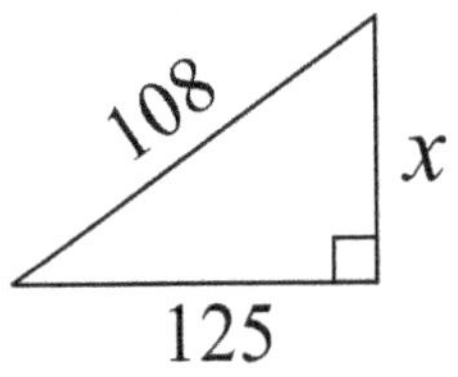
108
x
125

The Pythagorean Theorem: Applications

Example 1. An eight-foot ladder is placed against a wall so that the base of the ladder is 2 ft away from the wall. What is the height of the top of the ladder?

Since the ladder, the wall, and the ground form a right triangle, this problem is easily solved by using the Pythagorean Theorem. Let h be the unknown height. From the Pythagorean Theorem, we get:

$$2^2 + h^2 = 8^2$$
$$4 + h^2 = 64$$
$$h^2 = 60$$
$$h = \sqrt{60}$$
$$h \approx 7.75$$

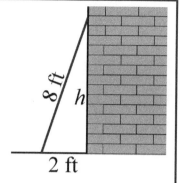

Our answer, 7.75, is in feet. This means the ladder reaches to about 7 3/4 ft = 7 ft 9 in. high.

1. Is this corner a right angle?

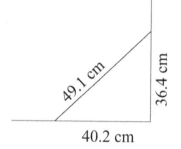

2. How long is the diagonal of a laptop screen that is 9.0 inches high and 14.4 inches wide?

   Note: when a laptop is advertised as having a "15-inch screen," it is the diagonal that is 15 inches, not the width or the height.

3. A park is in the shape of a rectangle and measures 48 m by 30 m. How much longer is it to walk from A to B around the park than to walk through the park along the diagonal path?

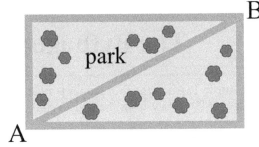

4. The area of a square is 100 m². How long is the diagonal of the square?

5. A clothesline is suspended between two apartment buildings. Calculate its length, assuming it is straight and doesn't sag any.

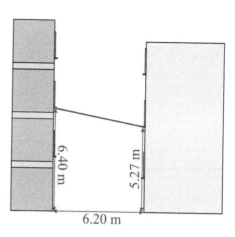

6. Construction workers have made a rectangular mold out of wood, and they are getting ready to pour cement into it. How could they make sure that the mold is indeed a rectangle and not a parallelogram? After all, in a parallelogram the opposite sides are equal, so simply measuring the opposite sides does not guarantee that a shape is a rectangle.

| **Example 2.** Find the area of this isosceles triangle. 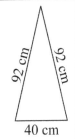 40 cm | **Solution:** To calculate the area of any triangle, we need to know its altitude. When we draw the altitude, we get a right triangle:

The next step is to apply the Pythagorean Theorem to solve for the altitude h, and after that calculate the actual area. 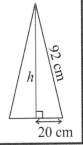 20 cm |

7. Calculate the area of the isosceles triangle in the example above to the nearest ten square centimeters.

8. Calculate the area of an equilateral triangle with 24-cm sides to the nearest square centimeter. Don't forget to draw a sketch.

9. Calculate the length of the rafter in feet and inches, if...

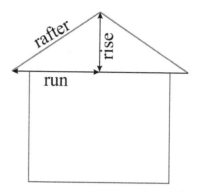

   a. ...the run is 12 ft and the rise is 3 ft

   b. ...the run is 12 ft and the rise is 5 ft 3 in.

10. Find the surface area of this roof to the nearest tenth of a square meter.

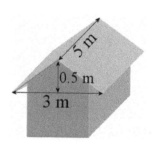

11. A creek runs through a piece of land in a straight line.

   a. Find the length of the creek. Give your answer to the same accuracy as the dimensions in the picture.

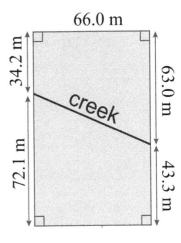

   b. The creek splits the plot into two parts. Calculate the areas of the two parts to the nearest ten square meters.

Puzzle Corner

The roof of a little kiosk is in the shape of a square pyramid. Each bottom edge measures 3.5 m, and the other edges measure 2.2 m. Calculate the surface area of this roof to the nearest tenth of a square meter.

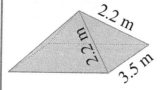

A Proof of the Pythagorean Theorem

There are hundreds of different proofs for the Pythagorean Theorem. In this lesson, we will look at two simple ones that are based on geometry.

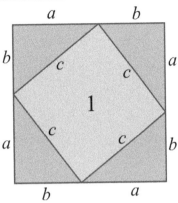

The figure above has four right triangles, each with sides a, b and c. The sides of the outside square are $a + b$. The triangles enclose a square with sides c units long.

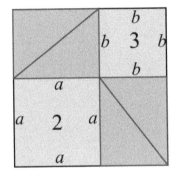

Here the sides of the large square are still $a + b$, but the four right triangles have been rearranged to form two smaller squares, with sides a and b.

Since the areas of both large squares are equal, and the areas of the four right triangles are equal, it follows that the remaining (blue) areas are also equal. In other words, the area of square 1, which is c^2, equals the area of square 2 (which is a^2) plus the area of square 3 (which is b^2). In symbols, it is $c^2 = a^2 + b^2$. ☺

1. Figure out how this proof of the Pythagorean Theorem works.

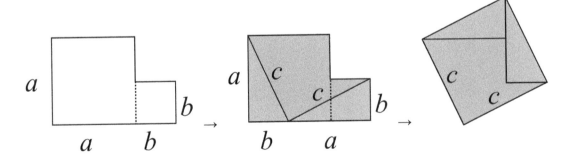

Chapter 9 Mixed Review

1. A farmer plowed three acres of his field, which is 2/5 of the total area he is planting this year. How many acres is he planting this year?

   a. Choose a variable for the unknown quantity, and write an equation for the problem.

   b. Solve the equation.

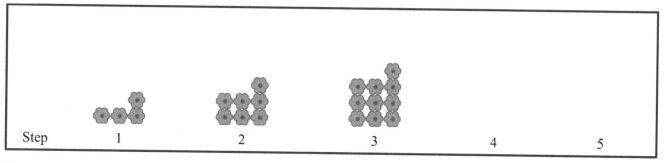

2. a. Draw the next two steps.

   b. How do you see this pattern growing?

   c. How many flowers will there be in step 39?

   d. What about in step n?

3. Fill in the missing parts in this justification for the rule *"Negative times negative makes positive."*

   (1) Substitute $a = -1$, $b = 1$, and $c = -1$ in the distributive property $a(b + c) = ab + ac$.

   ____ (____ + ____) = ____ · ____ + ____ · ____

   (2) The whole left side is zero because ____ + ____ = 0.

   (3) So the right side must equal zero as well.

   (4) On the right side, $-1 \cdot 1$ equals ____. Therefore, $-1 \cdot (-1)$ must equal ____ so that the sum on the whole right side will equal zero.

4. An educational website currently offers a subscription plan with a $14.95 monthly cost. The company is considering increasing it to $19.95. What is the percent of increase?

5. **a.** Write an equation for angle x and solve it.

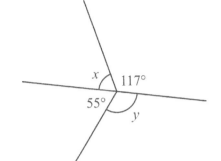

   b. Write an equation for angle y and solve it.

6. Calculate the area of a rectangle with sides 2 yards and 5 yards both in square yards and in square feet.

   A = _____ sq. yd. = _____ sq. ft.

7. Jane has a number of identical small cylinder-shaped water glasses. Their bottom diameter measures 6 cm, and their height is 8 cm. Jane fills them 3/4 full. How many glasses can Jane fill out of a 1-liter pitcher of juice? *Hint: 1 cm^3 = 1 ml.*

8. **a.** Draw a parallelogram that has 36° and 144° angles and sides that are 7 cm and 4.5 cm long.

   b. Find the area of the parallelogram.

9. A farmer gets paid $3 for a bushel of corn. A bushel is about 35.2 liters. How much would the farmer be paid for 200 liters of corn?

10. Angela and Mary solved the following problems differently.

   a. Who got each question correct?

   b. Which package of granola costs less by weight?

   A package of granola weighs 800 g and costs $3.60. Another package weighs 600 g and costs $3.00.
   (1) How many percent heavier is the first than the second?
   (2) How many percent more expensive is the first than the second?

   Angela: (1) I subtract 800 g − 600 g = 200 g and write the fraction $\frac{200}{800} = \frac{1}{4} = 25\%$.

   Angela: (2) I subtract $3.60 − $3.00 = $0.60 and write the fraction $\frac{0.60}{3.60} = \frac{1}{6} \approx 17\%$.

   Mary: (1) I subtract 800 g − 600 g = 200 g and write the fraction $\frac{200}{600} = \frac{1}{3} \approx 33\%$.

   Mary: (2) I subtract $3.60 − $3.00 = $0.60 and write the fraction $\frac{0.60}{3.00} = \frac{1}{5} = 20\%$.

11. Calculate.

   a. $8\frac{3}{4} \div 2\frac{5}{8} + 2\frac{2}{5}$

   b. $4\frac{2}{7} \div \frac{6}{7} \cdot \frac{5}{8}$

Chapter 9 Review

1. Find the square roots.

a. $\sqrt{144}$	b. $-\sqrt{81}$	c. $\sqrt{1,600}$
d. $\sqrt{10^2 - 6^2}$	e. $\sqrt{49 \cdot 49}$	f. $\sqrt{5 \cdot (83 - 3)}$

2. **a.** If the side of a square measures $\sqrt{7}$ cm, what is its area?

   b. How long is the side of a square with an area of 20 cm²?

3. Solve. Give your answer to the nearest thousandth. You may use a calculator.

a. $y^2 + 18 = 35$	b. $0.6h^2 = 4$

4. For each set of lengths, determine whether they form a right triangle.

   a. 20, 24, 30

   b. 2.6, 1.0, 2.4

5. Solve for the unknown side of each triangle. Remember, you can ignore the negative answer. (Why?)

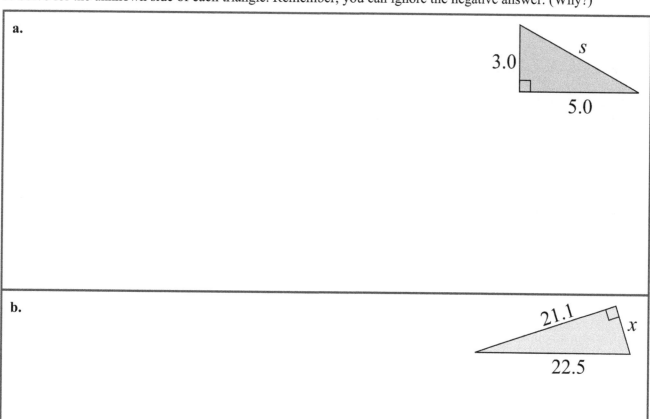

6. Lauren and Anna want to make this pennant for their jogging club. Calculate its total area.

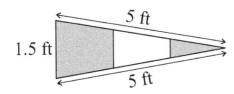

7. The map shows part of downtown Nashville, Tennessee. The triangle ABC on the map is very close to a right triangle. The distance AB is 370 m and the distance AC is 620 m. However, these distances are approximate, so your calculations will also be only approximate.

About how much shorter is it to travel from point A to point C along Lafayette Street than to travel first along Korean Veterans Boulevard and then along 5th Avenue South?

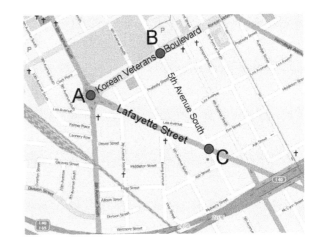

Chapter 10: Probability
Introduction

Probability is new to Math Mammoth students, as the topic doesn't appear at all in grades 1-6. However, most students have an intuitive understanding of probability based on hearing the terms "probably" and "likely," listening to weather forecasts, and so on.

In the past, probability wasn't taught until high school—for example, I personally encountered it for the first time in 12th grade. However, since probability is such a useful and easily accessible field of math, it was felt that it should be introduced sooner, so during the 1990s and 2000s it "crept" down the grade levels until many states required probability even in elementary school. The Common Core Standards include probability starting in 7th grade. I feel that is good timing because by 7th grade students have studied fractions, ratios, and proportions, so they have the tools they need to study probability. Moreover, they will need an understanding of the basic concepts of probability in order to understand the statistical concepts that they will study in middle school and high school.

In this chapter we start with the concept of simple (classic) probability, which is defined as the ratio of the number of favorable outcomes to the number of all possible outcomes. Students calculate probabilities that involve common experiments, to include flipping a coin, tossing a pair of dice, picking marbles, or spinning a spinner.

The lesson *Probability Problems from Statistics* introduces probability questions involving the phrase "at least," which are often solved by finding the probability of the complement event. For example, it might be easier to count the number of students who got at most D+ on a test than to count the number of students who got at least C-.

In the next lesson, *Experimental Probability*, students conduct experiments, record the outcomes, and calculate both the theoretical and experimental probabilities of events, in order to compare the two. They will draw a card from a deck or roll a die hundreds of times. The download version of this curriculum includes spreadsheet files for some of the lengthier probability simulations. You can also access those simulations at the web page http://www.mathmammoth.com/lessons/probability_simulations.php

Next, we study compound events, which combine two or more individual simple events. Tossing a die twice or choosing first a girl then a boy from a group of people are compound events. Students calculate the probabilities of compound events by using the complete sample space (a list of all possible outcomes). They construct the sample space in several ways: by drawing a tree diagram, by making a table, or simply by using logical thinking to list all the possible outcomes.

The last major topic in this chapter is simulations. Students design simulations to find the probabilities of events. For example, we let heads represent "female" and tails represent "male," so we can toss a coin to simulate the probability of choosing a person of either sex at random. Later in the lesson, students design simulations that use random numbers. They generate those numbers by using either the free tool at http://www.random.org/integers or a spreadsheet program on a computer.

In the last lesson of the chapter, *Probabilities of Compound Events*, we learn to calculate the probability of a compound event by *multiplying* the probabilities of the individual events (assuming the outcomes of the individual events are independent of each other). This topic exceeds the Common Core Standards for 7th grade and thus is optional. I have included it here because the idea studied in the lesson is very simple and I feel many students will enjoy it.

You can find matching videos for some of the topics in this chapter at
http://www.mathmammoth.com/videos/probability/probability_lessons.php

The Lessons in Chapter 10

	page	span
Probability	222	*3 pages*
Probability Problems from Statistics	225	*2 pages*
Experimental Probability	227	*3 pages*
Count the Possibilities	230	*6 pages*
Using Simulations to Find Probabilities	236	*6 pages*
Probabilities of Compound Events	242	*4 pages*
Chapter 10 Mixed Review	246	*3 pages*
Chapter 10 Review	249	*2 pages*

Helpful Resources on the Internet

Probability Videos by Maria
These video lessons cover topics that have been chosen to complement the lessons in this chapter.
http://www.mathmammoth.com/videos/probability/probability_lessons.php

SIMPLE PROBABILITY

Probability Game with Coco
A multiple-choice online quiz on simple probability.
http://www.math-play.com/Probability-Game.html

Probability Circus
Choose the spinner that matches the probability in this interactive online activity.
http://www.hbschool.com/activity/probability_circus/

Simple Probability
Practice finding probabilities of events, such as rolling dice, drawing marbles out of a bag, and spinning spinners.
https://www.khanacademy.org/math/cc-seventh-grade-math/cc-7th-probability-statistics/cc-7th-basic-prob/e/probability_1

Card Sharks Game
Use your knowledge of probability to bet on whether the next card is higher or lower than the last one.
http://mrnussbaum.com/cardsharks/

Mystery Spinners
In this activity, you need to find the hidden mystery spinner working from only one clue.
http://www.scootle.edu.au/ec/viewing/L2384/index.html

Simple Probability Quiz
Reinforce your probability skills with this interactive self-check quiz.
http://www.phschool.com/webcodes10/index.cfm?wcprefix=ara&wcsuffix=1201

EXPERIMENTAL PROBABILITY

Probability Tools
Play around with dice, coins, spinners, playing cards, counters and digit cards. In each type of game, you can choose from a variety of settings such as number of dice, number of trials, whether to display results as a frequency diagram or table.
http://www.interactive-maths.com/probability-tools-flash.html

Adjustable Spinner
Create a virtual spinner with variable-sized sectors to compare experimental results to theoretical probabilities. You can choose the sizes of the sectors, the number of sectors, and the number of trials.
http://www.shodor.org/interactivate/activities/AdjustableSpinner/

Interactive Customizable Spinners
Use a tool to build colored spinners. Then, test the spinner over a number of spins. Compare the actual results with the expected results.
https://fuse.education.vic.gov.au/Resource/LandingPage?ObjectId=8eb446b6-bd1e-446a-848a-935dce8b0b70

Dice Duels Tool
Explore the experimental probability distributions when you roll between two and five dice, and either add, subtract, or multiply the numbers. The tool graphs the results, and can do up to 9,999 rolls.
https://fuse.education.vic.gov.au/Resource/LandingPage?ObjectId=f8341288-4733-4604-abf3-7c1d9de7fc4b

Dice Roll
Choose the number of virtual dice to roll and how many times you want to roll them. The page shows both the actual results and expected (theoretical) probabilities, and the simulation works for a very large numbers of rolls.
http://www.btwaters.com/probab/dice/dicemain3D.html

Coin Flip
This virtual coin toss shows the results numerically and can generate at least 100,000 flips.
http://www.btwaters.com/probab/flip/coinmainD.html

Coin Toss Simulation
Another virtual coin toss. This one shows the results both using images of coins and numerically.
http://syzygy.virtualave.net/multicointoss.htm

Rocket Launch
A three-stage rocket is about to be launched. In order for a successful launch to occur, all three stages of the rocket must successfully pass their pre-takeoff tests. By default, each stage has a 50% chance of success, however, this can be altered by dragging the bar next to each stage. Observe how many tries it takes until there is a successful launch.
http://mste.illinois.edu/activity/rocket/

Find the Bias
A die (one cube of dice) has been weighted (loaded) to favor one of the six numbers. Roll the die to work out which is the favored face. Explore how many rolls are needed for you to be reasonably sure of a conclusion.
https://fuse.education.vic.gov.au/Resource/LandingPage?ObjectId=0ec1fe96-7c91-4975-9863-766a1fe9c1c5

Load One Dice (Biased dice)
Make a biased die by loading it to favor one of the six numbers. Then roll the loaded die, and compare the shape of theoretical data distributions with experimental results.
https://fuse.education.vic.gov.au/Resource/LandingPage?ObjectId=b60d5135-608d-4fd6-8e65-a81fa1dc172a

Racing Game with One Die
Explore how experimental probability relates to fair and unfair games with this two-car race. You choose which and how many numbers of the die make each of the cars move. Other options include the number of trials and the length of the race (in segments). The program calculates the percentage of wins for each car and draws a pie chart.
http://www.shodor.org/interactivate/activities/RacingGameWithOneDie/

COMPOUND EVENTS AND TREE DIAGRAMS

Tree Diagrams and Spinners Quiz
Practice reading tree diagrams in this interactive 10-question quiz.
https://www.thatquiz.org/tq/practicetest?8x24p16y3o5t

Lucky 16 Game
You place counters on the game board, and then they will be removed based on the sum of two dice that are rolled. Try to predict the best positions for the counters before the game starts.
https://fuse.education.vic.gov.au/Resource/LandingPage?ObjectId=31ee684d-742e-4fe3-bb55-e8e6e61e0d6d

Airport Subtraction Game
This game is based on rolling two dice and subtracting the results. You task is to place your plane at the back of the queue on one of the runways. Try to predict which lane is most likely to clear quickly.
https://fuse.education.vic.gov.au/Resource/LandingPage?ObjectId=0c1b063f-52ec-4564-ac6e-265ddcbdcada

Quiz: Compound Probability
Test your knowledge of compound probability with this interactive self-check quiz.
http://www.phschool.com/webcodes10/index.cfm?wcprefix=aba&wcsuffix=1205

How could I send the check and not pay the bill?
What is the probability that Tessellation will put each of the three checks into the correct envelopes if she does it randomly? The page includes a hint and a complete solution (click "answer" at the bottom of page).
http://figurethis.nctm.org/challenges/c69/challenge.htm

Flippin' Discs
In this interactive activity, you throw two discs. You win if they both show the same color. You can run the game 100 times and see the detailed results. Can you explain why you win approximately half the time? Explore the situation also with 3, 4, and even 5 discs. The solution is found in a link near the top left of the page.
http://nrich.maths.org/4304

"Data Analysis & Probability Games" from MathWire
A list of board and dice games to help to teach topics appropriate for beginners in probability.
http://mathwire.com/games/datagames.html

Cross the Bridge
This is a printable board game based on throwing two dice and the probabilities for the sum of the dice.
http://www.mathsphere.co.uk/downloads/board-games/board-game-17-crossing-the-river.pdf

At Least One…
The tree diagram and related discussion on this page guides students' thinking to help them answer probability questions like, "What is the probability of getting at least one head by flipping a coin ten times?" A link near the top left of the page leads to the solution.
http://nrich.maths.org/7286

SIMULATIONS

Probability Simulations in Excel
These spreadsheet files match some of the lengthier probability simulations in this chapter.
http://www.mathmammoth.com/lessons/probability_simulations.php

Random Integer Generator
Choose how many numbers, how many columns, and the values of the integers and then click to generate.
https://www.random.org/integers/

Coin Toss Simulation
Another virtual coin toss. This one shows the results both using images of coins and numerically.
http://syzygy.virtualave.net/multicointoss.htm

Marbles
Run repeated experiments where you draw 1, 2, or 3 marbles from a set of blue, red, purple, and green marbles. The results shown include the frequencies of each possible outcome, the experimental probabilities, and the theoretical probabilities. This activity can be used not only to explore probabilities but also to perform simulations.
http://www.shodor.org/interactivate/activities/Marbles/

Interactivate: *Fire!!* and *Directable Fire!!*
In these two activities, you first set the probability that a fire will spread from tree to tree in a forest of 100 trees. Then you click the tree where the fire starts and watch it spread. In the Directable Fire activity, you can set the probabilities for each direction to be different. Repeat the activity several times to see that the amount of forest that burns varies (for any set probability of fire spreading).
http://www.shodor.org/interactivate/activities/Fire/
http://www.shodor.org/interactivate/activities/DirectableFire/

COMPOUND PROBABILITY

Probability Quiz
Test your knowledge about probability in this interactive self-check quiz.
http://www.phschool.com/webcodes10/index.cfm?wcprefix=ama&wcsuffix=7454

Probability of Compound Events Quiz
Practice finding the probability of compound events. Some problems deal with replacement and non-replacement.
https://maisonetmath.com/probability/quizzes/409-probability-of-compound-events

Probability Quiz
Reinforce your skills with this interactive multiple-choice quiz.
http://www.phschool.com/webcodes10/index.cfm?wcprefix=aqa&wcsuffix=1005&area=view

FOR FUN

Monty Hall Paradox
Try this interactive version of the famous Monty Hall problem. Behind which door is the car? If you choose the wrong one, you'll win a goat instead.
http://www.math.ucsd.edu/~crypto/Monty/monty.html

What Does Random Look Like?
This problem challenges our thinking about randomness. Make up a sequence of twenty Hs and Ts that *could* represent a sequence of heads and tails generated by a fair coin. Then use the animation to generate truly random sequences of 20 coin flips. Can you learn how to spot fakes?
http://nrich.maths.org/7250

Same Number
Imagine you are in a class of thirty students. The teacher asks everyone to secretly write down a whole number between 1 and 225. How likely is it for everyone's numbers to be different? The web page provides an interactive simulation so you can experiment with this problem. The following discussion also leads students to the classic birthday problem. The solution is found in a link near the top left of the page.
http://nrich.maths.org/7221

Probability

You *probably* already have an intuitive idea of what *probability* is. In this lesson we look at some simple examples in order to study probability from a mathematical point of view.

If we flip a coin, the chance, or **probability**, of getting "heads" is 1/2, and the chance of getting "tails" is also 1/2. "Heads" and "tails" are the two possible **outcomes** when you toss a coin, and they are equally likely.

When rolling a six-sided number cube (a die), you have six possible **outcomes**: you can roll either 1, 2, 3, 4, 5, or 6. These are all equally likely (assuming the die is fair).

Thus the probability of rolling a five is 1/6. The probability of rolling a three is also 1/6. In fact, the probability of each of the six outcomes is 1/6.

The probability of rolling an even number is 3/6, or 1/2, because three of the six possible outcomes are even numbers.

Simple probability has to do with situations where each possible outcome is <u>equally likely</u>.

Then the **probability** of an event is the fraction $\dfrac{\text{number of favorable outcomes}}{\text{number of possible outcomes}}$

"Favorable outcomes" are those that make up the event you want. The examples will make this clear.

Example 1. What is the probability of getting a number that is less than 6 when tossing a fair die?

Count how many of the outcomes are "favorable" (less than 6). There are five: 1, 2, 3, 4, or 5.
And there are six possible outcomes in total.

Therefore, the probability is $\dfrac{\text{number of favorable outcomes}}{\text{number of possible outcomes}} = \dfrac{5}{6}$.

In math notation we write "P" for probability and put the event in parentheses:

P(less than 6) = 5/6.

Example 2. On this spinner the number of possible outcomes is eight, because the arrow is equally likely to land on any of the eight wedges. What is the probability of spinning yellow?

There are TWO favorable outcomes (yellow areas) out of EIGHT possible outcomes (wedges).

P(yellow) = 2/8 = 1/4.

(Because green and yellow each have two wedges, there are only six possible colors that can result. When we list the possible outcomes, we list the six colors. However, when we figure the probabilities, we must use the eight equal-sized wedges to find the probability.)

By convention, the probability of an event is always at least 0 and at most 1. In symbols: $0 \leq P(\text{event}) \leq 1$.

A probability of 0 means that the event does not occur; it is impossible. Probability of 1 means that the event is sure to occur; it is certain. A probability near 1 (such as 0.85) means that the event is likely to occur. A probability of 1/2 means that an event is neither likely nor unlikely.

Example 3. What is the probability of rolling 8 on a standard six-sided die?

This is an impossible event, so its probability is zero: P(8) = 0.

Example 4. What is the probability of rolling a whole number on a die?

This is a sure event, so its probability is one. P(whole number) = 1.

1. There are three red marbles, two dark blue marbles, and five light green marbles in Michelle's bag. List all the possible outcomes if you choose one marble randomly from her bag.

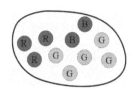

2. Michelle chooses one marble at random from her bag. What is the probability that...

   a. the marble is blue?

   b. the marble is not red?

   c. the marble is neither blue nor green?

3. Make up an event with a probability of zero in this situation.

4. Suppose you choose one letter randomly from the word "PROBABILITY."

   a. List all the possible outcomes for this event.

   Now find the probabilities of these events:

   b. P(B)

   c. P(A or I)

   d. P(vowel)

   e. Make up an event for this situation that is likely to occur, yet not a sure event, and calculate its probability.

> **The complement of an event and the probability of "not"**
> The **complement** of any event A is the event that A does *not* occur.
> If the probability of event A is a, then the probability of A not happening is simply $1 - a$.

5. The weatherman says that the chance of rain for tomorrow is 1/10. What is the probability of it not raining?

6. The spinner is spun once. Find the probabilities as simplified fractions.

   a. P(green)   b. P(not green)

   c. P(not pink) d. P(not black)

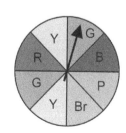

   e. Make up an event for this situation that is not likely, yet not impossible either, and calculate its probability.

223

Probabilities are often given as percentages instead of fractions.

Example 5. Kimberly's sock bin contains 7 brown socks, 9 white socks, and 5 red socks. She picks one without looking. What is the probability that she gets a white sock?

There are 9 white socks out of 21 socks in all. The probability is 9/21 = 3/7 = 3 ÷ 7 ≈ 0.42857 = 0.429 = 42.9%.

7. Suppose you were to draw one card from the set of cards on the right. Complete the table with the possible outcomes, and their probabilities both as fractions and as percentages (to the nearest tenth of a percent).

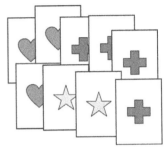

Possible outcomes	Probability (fraction)	Probability (percentage)

8. This "rainbow spinner" is spun once. Find the probabilities to the nearest tenth of a percent.

   a. P(yellow)   **b.** P(blue or green)

   c. P(not orange)   **d.** P(not red and not purple)

   e. Make up an event for this situation with a probability of 1.

9. **a.** An empty bus has 45 seats, and 22 of them are window seats. If you are assigned a seat at random, what is the probability, to the nearest tenth of a percent, that you get a window seat?

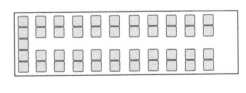

   b. Now each seat marked with an "x" is already occupied. If you choose a seat randomly, what is the probability, to the nearest tenth of a percent, that you get a window seat?

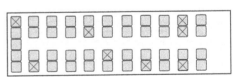

The chart shows you the seating arrangement of a bus. You enter the bus, and the driver informs you that fifteen seats are already occupied and that if you choose a seat randomly, the probability of getting a window seat is less than 25%.

How many window seats are occupied, at least?

Puzzle Corner

Probability Problems from Statistics

Example 1. The bar graph shows the science test scores of all seventy 7th graders in Westmont School. If you choose one of them at random, then what is the probability that the student's score was at least C− (in other words, C− or better)?

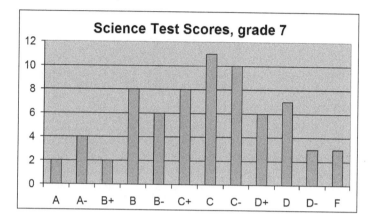

Sometimes when a probability question involves "at least," it is easier to look at the complement event — everything else — and find its probability first. The complement of "at least C−" is "at most D+" in other words, D+, D, D−, and F. From the graph, it is easier to sum the number of students who got the four low scores than to sum the number of students who got the eight high scores.

The number of students who got D+, D, D−, or F is 6 + 7 + 3 + 3 = 19 students. There are a total of 70 students, so P(at most D+) = 19/70. Now it's easy to calculate the original probability in question: P(at least C−) = 1 − 19/70 = 51/70.

1. You choose one student at random from the 7th graders in Westmont School.

    a. What is the probability that the student's score was at least D?

    b. What is the probability that the student's score was at most B+?

2. The dotplot shows the age distribution of a children's fishing club. One child is chosen randomly from the group.

    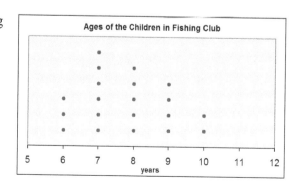

    a. What is the probability that the child is at most 9 years of age?

    b. What is the probability that the child is at least 7 years of age?

3. The chart lists the languages most commonly spoken at home in the United States and the number of people at least 5 years old who speak them.

   If one person is selected randomly from this population, what is the probability, to the nearest hundredth of a percent, that the person does *not* speak only English at home?

Only English	215,423,557
Spanish	28,101,052
Other Indo-European	10,017,989
Asian Language	6,960,065
Other	1,872,489
Total Population Age 5+	262,375,152

4. The table lists the numbers of males and females in various age groups in the United States in the year 2000. Answer the questions to the nearest tenth of a percent.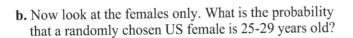

   a. What is the probability that a randomly chosen *person* in the United States is a female, 25-29 years old?

   b. Now look at the females only. What is the probability that a randomly chosen US female is 25-29 years old?

   c. What is the probability that a randomly chosen person in the United States is a male who is at least 15 years old?

   d. What is the probability that a randomly chosen person in the United States is at most 64 years old?

Age	Male	Female	Both sexes
0-4	9,810,733	9,365,065	19,175,798
5-9	10,523,277	10,026,228	20,549,505
10-14	10,520,197	10,007,875	20,528,072
15-19	10,391,004	9,828,886	20,219,890
20-24	9,687,814	9,276,187	18,964,001
25-29	9,798,760	9,582,576	19,381,336
30-34	10,321,769	10,188,619	20,510,388
35-39	11,318,696	11,387,968	22,706,664
40-44	11,129,102	11,312,761	22,441,863
45-49	9,889,506	10,202,898	20,092,404
50-54	8,607,724	8,977,824	17,585,548
55-59	6,508,729	6,960,508	13,469,237
60-64	5,136,627	5,668,820	10,805,447
65-69	4,400,362	5,133,183	9,533,545
70-74	3,902,912	4,954,529	8,857,441
75-79	3,044,456	4,371,357	7,415,813
80-84	1,834,897	3,110,470	4,945,367
85+	1,226,998	3,012,589	4,239,587
Totals	138,053,563	143,368,343	281,421,906

Experimental Probability

> In this lesson, we study **experimental probability**, which refers to the probability of an event based on actually conducting an experiment and observing how often the event occurs. It is the ratio of the number of times an event occurred to the number of times tested.
>
> This is in contrast to **theoretical probability**, which is calculated using mathematics and logical thinking, without actually conducting any experiments.

1. In this exercise, you will roll a die a lot of times to explore whether the chance of getting 1, 2, 3, 4, 5, and 6 is indeed 1/6 like it is theoretically.

    a. You will be rolling a (fair) die 60 times. Based on the theoretical probabilities, we would expect that each of the numbers 1, 2, 3, 4, 5, and 6 would come up exactly 1/6 · 60 = 10 times. Will that happen in reality?

    Roll a die 60 times and record each outcome. Count the **frequency** (how many times) that each number occurred and list the results in the table below. In the last column, calculate the percentages of how often each outcome occurred, to the hundredth of a percent.

    b. Now you will be rolling a die 120 times. How many times would you expect to roll each number, based on the theoretical probabilities of 1/6? _____ times

    However, we know that that will probably not happen. We can predict that each number will be rolled roughly that many times, but probably not exactly that many.

    Roll a die 120 times (start your count anew), record the outcomes again, and fill in the table for part (b). (Use multiple dice and several persons rolling them for quicker results.)

    c. One more time: roll the die 480 times, record the outcomes, and fill in the table for part (c).

a. With 60 rolls:			b. With 120 rolls:			c. With 480 rolls:		
Outcome	Frequency	Probability	Outcome	Frequency	Probability	Outcome	Frequency	Probability
1			1			1		
2			2			2		
3			3			3		
4			4			4		
5			5			5		
6			6			6		
TOTALS	60	100%	TOTALS	120	100%	TOTALS	480	100%

   d. Theoretically, the probability of rolling any of the numbers is 1/6 or 16.67%. In which of the three experiments — rolling the die 60 times, 120 times, or 480 times — were the experimental probabilities closest to the theoretical?

227

2. Through the marvels of automation, you will now "roll" a die even more times than in Exercise 1. Use the virtual dice roller at http://www.btwaters.com/probab/dice/dicemain3D.html or if you have the download version of the curriculum, you can run the simulation in the included spreadsheet file.

   Note: If you roll two or more dice in this simulation, the results show the sum of the dots on the dice, not the actual numbers that were rolled. So you'll want to leave the input value for "number of dice" set to "1."

   a. Predict about how many times you expect to get each of the six possible numbers if you roll a die 1,000 times:

   About _____ times

   b. Now roll one die 1,000 times.

   If you use the virtual roller, choose "session" (and not "historical") to see the data for your session. To rerun the simulation you need to refresh the page (press F5).

   Record in the table the frequencies of each outcome and calculate experimental probabilities.
   Observe how close each experimental probability is to the theoretical probability of 1/6 = 16.67%.

   c. Let's say you were to roll a die 5,000 times. How do you expect the results to differ from rolling a dice 1,000 times?

Outcome	Frequency	Experimental Probability (%)
1		
2		
3		
4		
5		
6		
TOTALS	1,000	100%

   d. (Optional) Try some larger simulations to see if you can determine about how many rolls it takes for the experimental probabilities to fall within half a percentage point of the theoretical value (from 16.17% to 17.17%)?

3. Select a set of 12 cards from a deck of playing cards so that the set has five different kinds of cards in it. Choose numbers from different suits. For example, you could choose the cards 2, 2, 2, 3, 3, 4, 5, 5, 5, 5, 6, and 6. The experiment will involve choosing a card at random from your set.

   a. Choose a card at random from your set, record the outcome, and put the card back. Repeat this 100 times. Count the frequencies of each outcome. In the table, record the **relative frequencies**—the frequencies written as fractions of the total number of repetitions.

   b. Calculate the theoretical and experimental probabilities for each outcome. Then compare the two: are they fairly close?

   If not, what could have caused the discrepancy?

Outcome	Theoretical probability	Relative Frequency	Experimental probability
	____%	____/100	____%

   You can also run this simulation at http://nlvm.usu.edu/en/nav/frames_asid_146_g_3_t_5.html
   First, select which number cards the simulation will draw from. Tick the box for "Quick draw" and enter a number of repetitions. Press "Start." A bar graph shows the relative frequencies in your experiment. Click on the bars to see the relative frequencies as decimals.

4. You will now conduct an experiment where the various outcomes are not equally likely to occur. In such a case, we say that the probability model is **not uniform**. Choose from one of the experiments listed or come up with one of your own. Repeat the experiment 100 times and count how many times each outcome occurs. Then calculate the experimental probabilities.

(1) Toss a paper cup and observe how it lands: open-end down, open-end up, or on its side.

(2) Spin a coin and observe how it stops: heads up, tails up, or on its side.

(3) Roll a number cube that is slightly weighted on one side.

(4) Choose one card randomly from a deck of cards where some cards are sticking to each other. Observe whether you get a diamond. Put the card back into the deck in a random location before choosing the next card.

(5) Choose one card randomly from a deck of cards and observe whether you get a diamond. If it is *not* a diamond, put the card back randomly into the deck. If it *is* a diamond, place the card at the *back* of the deck. Don't reshuffle the deck at any point.

(6) Put several stuffed animals in a hat or box. The animals should vary in size as much as possible. Each time, pull out one animal randomly, then put it back.

If you would like, or if your teacher so decides, you could do more than one of these experiments.

Outcome	Relative Frequency	Experimental probability (%)
	____/100	

Counting the Possibilities

A **sample space** is a list of all possible outcomes of an experiment.

Example 1. We roll two dice. The sample space for this experiment is shown in the grid on the right. Each dot represents one outcome. For example, the point (1, 4) means that the first die shows 1 and the second die shows 4.

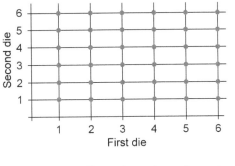

Notice that there are a total of 6 · 6 = 36 possible outcomes.

What is the probability of getting the sum of 8 when rolling two dice? The chart helps answer that question. First we find out and count how many outcomes give you the sum 8:

You could roll 2 + 6, 3 + 5, 4 + 4, 5 + 3, or 6 + 2. Those number pairs are circled in the second graphic.

So there are five favorable outcomes out of 36 possible. Therefore, the probability of getting 8 as a sum is 5/36.

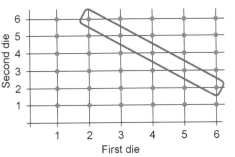

1. **a.** How many outcomes are there for rolling the same number on both dice (such as (5, 5))?

   b. What is the probability of rolling the same number on both dice?

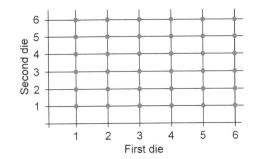

2. **a.** What is the probability of rolling 5 on the first die and 6 on the second?

   b. What is the probability of rolling 5 on one die and 6 on the other?

   c. What is the probability of getting a sum of 7 when rolling two dice?

   d. What is the probability of getting a sum of at least 6 when rolling two dice?

3. You roll a six-sided die twice. Find the probabilities.

   a. P(1; 5)

   b. P(2; 5 or 6)

   c. P(even; odd)

   d. P(6; not 6)

The array we used on the previous page can show the sample space (all the possible outcomes) for only two events, like rolling two different dice. A **tree diagram** can show more than two events, so it is a common way to represent the sample space for multiple events.

Example 2. Peter has white, blue, yellow, and red shirts, blue and white slacks, and brown and blue tennis shoes. How many possible ways can he make an outfit using them?

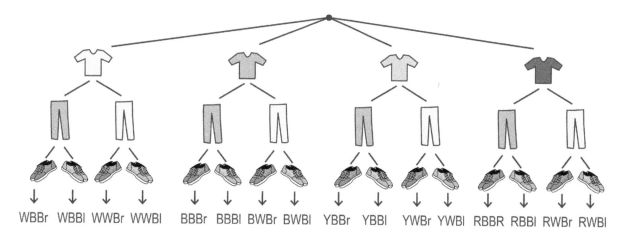

At the bottom we have listed all the possible outcomes using letter combinations. This is optional, but helpful. For example, WBBr means a white shirt, blue slacks, and brown shoes.

Notice that in the first level, there are 4 possibilities, in the second level there are 2 possibilities, and in the last level 2 possibilities. In total, there are $4 \times 2 \times 2 = 16$ ways he can make an outfit.

Example 3. Peter chooses his shirt, slacks, and shoes randomly. What is the probability that his shirt and slacks match?

"Matching" means that he wears a white shirt with white pants or a blue shirt with blue pants. Since the shoes are not specified, there are four possible outfits: WWBr, WWBl, BBBr, or BBBl. So the probability is 4/16 = 1/4.

Example 4. What is the probability that Peter wears a red shirt?

You can count the four outfits with a red shirt to get the probability as 4/16 = 1/4. But it's simpler to ignore the other clothing items and just look at the shirts: the red shirt is one of the four, so the probability is 1/4.

4. **a.** Complete the tree diagram to show the outcomes when you first roll a die, then toss a coin. The bottom row lists the outcomes using number-letter combinations, such as 1H and 1T.

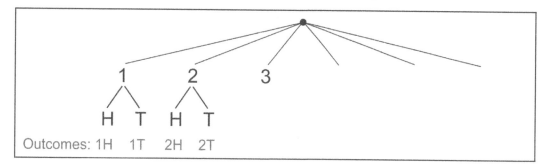

Now find these probabilities:

b. P(even number, heads)

c. P(not 6, heads)

d. P(4 or more, tails)

e. P(any number, tails)

5. A restaurant offers the following menu:

   Second course: soup or salad

   Main course: fish, chicken, or beef

   Dessert: ice cream or cake

   a. Cindy chooses her second course, main course, and dessert randomly. Draw a tree diagram for the sample space.

   What is the probability that Cindy…

   b. … gets ice cream for dessert?

   c. … gets soup, fish, and cake?

   d. … eats soup and fish (and either dessert)?

   e. … eats salad and ice cream?

   f. … doesn't eat chicken?

   g. … doesn't eat fish or ice cream?

6. You take a marble out of the bag and *put it back*. Then you take another marble out. Complete the table that lists the sample space (all the possible outcomes). Notice that we have to list both red marbles and both green marbles separately.

   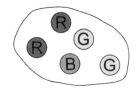

Second marble → First marble ↓	R	R	G	G	B
R	RR	RR	RG	RG	RB
R	RR				
G	GR				
G	GR				
B	BR				

   Now find the probabilities:

   a. P(red, then green)

   b. P(green, then red)

   c. P(not blue, not blue)

   d. P(not red, not red)

7. You take a marble out of this bag *without* putting it back, and then you take another marble. In effect, you take two marbles out of the bag.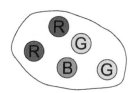

   a. Complete the tree diagram for this experiment. Notice: which marble you take out determines which marbles are left. For example, if the first marble is red, then the bag has 1 blue, 2 green, and *only* 1 red marble left to choose from.

   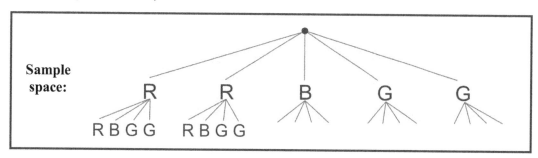

   Then find the probabilities:

   b. P(not red, not red)

   c. P(red, then green)

   d. P(green, then red)

   e. Add the probabilities from (c) and (d) to get the probability of choosing exactly one red and one green marble, in either order.

   f. (Optional) Conduct this experiment. If you don't have marbles, you could let red = quarters, green = dimes, and blue = nickels, and perform the experiment with coins. Observe for example whether the probability you calculated in (e) for getting one red and one green marble is close to what you observe in your experiment.

8. You make a two-digit number by choosing both digits randomly from the numbers on the cards. The card is replaced after each choice.

   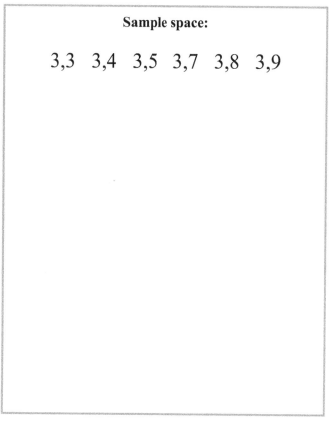

   Sample space:

   3,3   3,4   3,5   3,7   3,8   3,9

   a. In the space on the right, finish listing all the possible outcomes of this experiment.

   Use the list to find these probabilities:

   b. P(4; 9)

   c. P(even; 7)

   d. P(even; odd)

   e. P(less than 6; more than 6)

   f. P(not 6; not 6)

   g. P(both digits are the same)

9. A special education classroom has 4 boys and 2 girls. The teacher randomly chooses two students to be responsible for the cleanup after a bake sale.

   a. Make a tree diagram for the sample space. Notice that if the first student is a girl, then there are 4 boys and 1 girl left to choose the second student from. If the first student is a boy, then there are 3 boys and 2 girls left to choose the second student from.

   Now use the sample space and give these probabilities as fractions.

   b. What is the probability that both students are girls?

   c. What is the probability that both students are boys?

   d. What is the probability that the first student chosen is a girl, and the second is a boy?

   e. What is the probability that the first student chosen is a boy, and the second is a girl?

   Now check. The probabilities you get in (b), (c), (d), and (e) should total 1 because they are all the possible outcomes.

   f. Add the probabilities in (d) and (e) to get the probability that one of the cleaners is a girl and one is a boy.

10. In tossing two distinct coins, one of the possible outcomes is HT: first coin heads, second coin tails.

   a. List all the possible outcomes.

   b. Each of the possible outcomes is equally likely. Therefore, what is the theoretical probability of each outcome?

   c. Now toss two coins 200 times and compare the experimental probabilities to the theoretical ones. Before you do, predict about how many times you would expect to see each outcome:

   _____ times

   Note: You need to distinguish the coins somehow: Either use different coins, like a 5 c and a 10 c, or mark identical coins in some manner, maybe as "1" and "2." Distinguishing the coins is necessary because the outcomes HT and TH aren't the same. You need to know which is which.

   If you have the download version of the curriculum, you may also run the simulation in the included spreadsheet file.

   d. Check whether the observed frequencies are fairly close to those predicted by the theoretical probabilities.

   Let's say they were not. What could be the reason?

Outcome	Frequency	% of total tosses
TOTALS	50	100%

Puzzle Corner

In a multiple-choice test, you have four choices (a, b, c, and d) for your answer each time.

a. Let's say the test has two questions and Andy chooses both answers randomly. What is the probability that Andy gets both questions correct?

b. Let's say the test has *five* questions and Kimberly answers them all randomly. What is the probability she gets them all correct?

Using Simulations to Find Probabilities

1. Let's say that a reporter interviews ten people at random on the street. If the probability that he selects a male or a female is 50/50, then what is the probability that, of the ten people, exactly 5 are male and 5 are female?

   We will use a *simulation* to answer this question.

   If you have the download version of this curriculum, you can use the included spreadsheet file, which simulates choosing 10 people by generating sets of 10 random numbers (0s and 1s).

   Another possibility is to use the virtual coin tossing tool at http://syzygy.virtualave.net/multicointoss.htm Choose 10 coins and 1 set of coin tosses. Record the outcome (how many males/females). Repeat your experiment at least 100 times, recording each outcome.

   Yet a different way to do the simulation is to toss 10 coins 100 times yourself or with the help of others.

   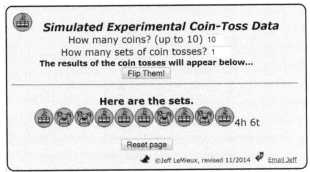

   http://syzygy.virtualave.net/multicointoss.htm

   A result of four heads and six tails simulates interviews with four females and six males.

   a. Fill in the table on the right based on the simulation.

   b. Based on your simulation, what is the experimental probability that exactly 5 of the randomly chosen people are female and 5 of them are male?

   c. What is the probability that 4 of the people are female and 6 are male?

   d. What is the probability of getting at least 3 females *and* at least 3 males in the set of 10 people?

   Hint: there are several possible outcomes with at least 3 females and at least 3 males.

Outcome	Frequency	Experimental probability
0 F 10 M		
1 F 9 M		
2 F 8 M		
TOTALS		100%

   e. What is the probability of getting only 1 or 2 of either sex (and 9 or 8 of the other)?

2. The chance that a random student at Lazyville High School has completed homework on time is 50%. One day, the principal chooses six of the school's students at random. Design a simulation to study this situation.

Explain your design:

Now run the simulation. Repeat the experiment at least 100 times, but more is better. Record each outcome. Then count how many of your outcomes represent zero students finishing homework on time, one student finishing it on time, and so on, and calculate the corresponding probabilities.

Results of simulation		
Students who finished homework	Relative Frequency	Experimental probability
0		
1		
2		
3		
4		
5		
6		
TOTALS		100%

Now use the results of the simulation to answer the following questions:

a. What is the probability that two of the six students have completed homework on time?

b. What is the probability that only one student has completed homework on time?

c. What is the probability that none of the six completed homework on time?

d. What is the probability that *at most* 2 of them have completed homework on time?
Hint: add the probabilities from (a), (b), and (c).

e. What is the probability that *at least* 3 of them have completed homework on time?

3. Let's say that the students at Lazyville High School have changed for the better and that the chance that a random student has completed homework on time is now 70%. Once again, the principal chooses six students at random. What is the probability that at least three of them have completed homework on time?

This time, we cannot use coin tosses to simulate this experiment because the probability that a single student has completed homework on time is no longer 50%, but we can still use simulation. Make up a deck of cards so that 70% of them are diamonds. For example, make a deck with 7 diamonds and 3 other cards. Then choosing one card randomly represents choosing a student randomly.

Now choose one card randomly from your deck and put it back. Repeat this six times. The six choices represent *one* outcome of the principal choosing six students). Record the outcome, for example as DXDDXD (D = diamond; X = other card).

Now repeat this experiment (choosing one card 6 times) at least 50 times (more is better). Record each outcome. Then count how many of your outcomes represent zero students finishing homework on time, one student finishing it on time, and so on, and calculate the corresponding experimental probabilities.

Students who finished homework	Relative Frequency	Experimental probability
0		
1		
2		
3		
4		
5		
6		
TOTALS		100%

Use the results of the simulation to answer the following questions:

a. What is the probability that at most 2 of them have completed homework on time?

b. What is the probability that at least 3 of them have completed homework on time?

c. What is the probability that not all of them have completed homework on time?

d. What is the probability that at least two-thirds of them have completed homework on time?

Example 1. Choosing one card from a deck hundreds of times, like you did in exercise 3, is a lot of work. An easier approach is to use computer-generated random digits, such as from **http://random.org/integers**

For the situation in exercise 3, you can generate integers between 0 and 9. To get the 70% probability, let the integers from 0 to 6 represent completing homework on time and 7 to 9 represent not completing homework on time.

Generating 600 integers and formatting the results in 6 columns will get you 100 rows with six numbers in each. Each row of six numbers represents one repetition of choosing of 6 students randomly.

> **Part 1: The Integers**
>
> Generate 600   random integers (maximum 10,000).
>
> Each integer should have a value between 0   and 9
>
> Format in 6   column(s).

For example, in the row 6 1 1 1 9 9, four out of the six numbers are 6 or less. This represents choosing a set of six students where four of them had completed homework on time.

If you don't have Internet access, you can create random numbers in a spreadsheet program. For example, in Excel, type the formula =FLOOR(RAND()*10, 1) into a cell.

RAND() creates a random decimal number between 0 and 1. Then we multiply it by 10 and round it down to the nearest whole number (the "floor") in order to get a random number between 0 and 9.

Lastly, to get a sequence of random numbers, copy that formula to a large number of cells.

4. Repeat exercise 3 using random digits. Generate at least 200 repetitions of choosing six students. Here is the situation again: The chance that a random student from Lazyville High School has completed homework on time is 70%. The principal chooses six students at random.

Students who finished homework	Relative Frequency	Experimental probability
0		
1		
2		
3		
4		
5		
6		
TOTALS		100%

a. What is the probability that at least half of them have completed homework on time?

b. What is the probability that at least four out of the six have completed homework on time?

5. A certain health care center typically gets 20 people per day who come to donate blood. Forty percent of them have type A blood. We are interested in determining how many donors we'll need to check before we find one with type A blood

   This question can be framed more exactly in terms of probabilities. For example, we could ask:

   - What is the probability that the first donor of the day has type A blood?
   - What is the probability that the first donor of the day *doesn't* have type A blood but the second one does? Let's denote this event as XA, where X means a person with some other blood type and A means a person with blood type A.
   - What is the probability of the event XXA — that only the third donor of the day has type A blood?
   - What is P(XXXA) — that it will take 4 donors to find one with type A blood?
   - What is P(XXXXA) — that it will take 5 donors to find one with type A blood?

   And so on.

   a. Design a simulation to study these types of questions. Explain your design here:

   b. Let's look at just the first four donors. Limit your simulation to a sequence of four donors, and repeat it at least 100 times. More is better. Record each outcome. Then count the frequencies of each listed event to complete the table.

   Note: in the table, the "_" denotes a person with any blood type, including type A. For example, the event "AXAA" is classified as part of the event "A _ _ _".

Results of simulation		
Event	Frequency	Experimental probability
A _ _ _		
XA _ _		
XXA _		
XXXA		
XXXX		
TOTALS		100%

   Find the probabilities:

   c. What is the probability that it will take 1, 2, or 3 donors until you find one with blood type A?

   d. What is the probability that it will take exactly 4 donors until you find one with blood type A?

   e. What is the probability that it will take more than 4 donors until you find one with blood type A?

   f. What is the probability that it will take at least 4 donors until you find one with blood type A?

6. An ideal gumball machine with an unending supply of gumballs dispenses them in four different flavors: strawberry, lemon, blackberry, and apple. Each flavor is equally likely.

   a. You get *three* gumballs out of the dispenser (randomly). Design a simulation for this experiment. Explain your design here:

   b. Run your simulation for at least 100 repetitions (but more is better). Record the outcomes.

   Hint: If you generate the random numbers at http://random.org/integers, copy the rows of random numbers to a spreadsheet program, such as Excel. You may need to use the "Paste special" command to paste them as "text." Then you can sort them easily. Simply choose all the cells you want to sort and then find "Sort" in the program's commands. In Excel it is "Data → Sort." Then choose the column by which to sort them. For example, if you want to study the first gumball in the set of three, sort the data by the first column.

   Find the experimental probabilities:

   c. P(none are strawberry) =

   d. P(exactly 2 are strawberry) =

   e. P(all 3 are strawberry) =

   f. P(none are lemon or blackberry) =

   g. P(all three are the same flavor) =

Probabilities of Compound Events

(This lesson is optional.)

Up to now we've been looking only at **simple events**, events that require just a single calculation of probability. A **compound event** is an event that consists of two or more simple events. If the outcome of one event does not affect the outcome of another, the events are said to be **independent**. If the compound event consists only of independent simple events, then it is very easy to calculate the probability of the compound event: we simply multiply the probabilities of the individual simple events. The examples will make this clear.

Example 1. You roll a die and toss a coin. What is the probability of rolling a 6 and getting heads?

P(6) is 1/6 and P(heads) is 1/2. Clearly, whether you get heads or tails on the coin does not affect what you get on the roll, so the two events are independent. Therefore, we can multiply the two probabilities:

$$P(6 \text{ and heads}) = \frac{1}{6} \cdot \frac{1}{2} = \frac{1}{12}$$

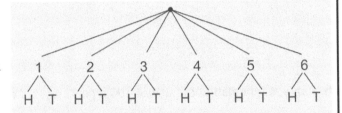

The tree diagram, too, shows that the probability of a 6 and heads is 1/12.

Example 2. You toss a coin three times. What is the probability of getting heads every time?

These three events—toss a coin, toss a coin, toss a coin—are independent. Getting heads on one toss doesn't affect whether you get heads or tails on the next.

P(heads) = 1/2. Therefore, $P(\text{heads and heads and heads}) = \frac{1}{2} \cdot \frac{1}{2} \cdot \frac{1}{2} = \frac{1}{8}$.

You can also see this result in the tree diagram. Only one outcome out of the 8 is "HHH."

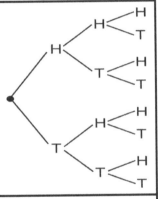

Example 3. A bag has three red marbles, two blue marbles, and five green marbles. You take out a marble and put it back. Then you take out a marble again and put it back. What is the probability of getting first a red marble and then a blue one?

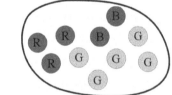

Again, we simply multiply the individual probabilities:

$$P(\text{red, blue}) = \frac{3}{10} \cdot \frac{2}{10} = \frac{6}{100} = \frac{3}{50}.$$

1. You toss a coin three times.

    a. What is the probability of getting tails, then heads, then tails?

    b. What is the probability that you get heads on your second toss?

    c. Use the tree diagram in Example 2. What is the probability of getting heads twice and tails once in three tosses? Note that they can be in any order, such as THH or HTH.

2. You take a marble out of the bag and put it back. Then you take out another marble. Find the probabilities:

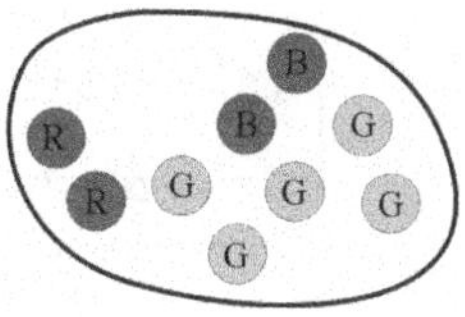

   a. P(red, then green)

   b. P(green, then red)

   c. P(not blue, not blue)

   d. P(not green, not green)

3. Helen is a teacher. She has eight different outfits that she wears to school, and each day she chooses her outfit randomly from among those eight. One of her outfits is pink, and her students don't like it.

   a. What is the probability that she does *not* wear the pink outfit for five days in a row?

   b. What is the probability that she does not wear the pink outfit for ten days in a row?

4. You make a two-digit number by choosing the digits randomly using these number cards. (You put the card back after choosing.)

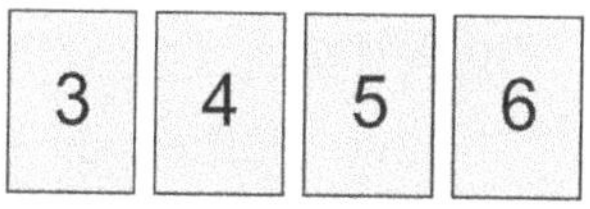

   a. How many different numbers is it possible to make?

   b. What is the probability of making the number 35?

   c. What is the probability of making a number that is divisible by 9?

5. The weatherman says that the chance of rain is 20% for each of the next five days, and your birthday is in two days! You also know that the probability of your dad taking you to the amusement park on your birthday is 1/2.

   a. What is the probability that you get to go to the park, and it doesn't rain?

   b. What is the probability that you get to go to the park, and it rains?

   Check: The sum of the probabilities in (a) and (b) should be 1/2.

Taking an object without replacing it

Example 4. You choose a marble from this bag and don't put it back. Then you choose another marble. This means you have in effect chosen two marbles. What is the probability that both are red?

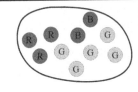

We first find the probability that the *first* marble is red. That is simply 3/10 since there are three red marbles and ten in all.

After you get a red marble and don't put it back, the bag now has only two red marbles and nine in total. So the probability of getting a red marble now is 2/9.

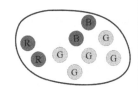

We multiply the two probabilities to find the probability of getting two red marbles:

$$P(\text{red and red}) = \frac{\overset{1}{\cancel{3}}}{\underset{5}{\cancel{10}}} \cdot \frac{\overset{1}{\cancel{2}}}{\underset{3}{\cancel{9}}} = \frac{1}{15}$$

6. You choose one card without putting it back. Then you choose another. What is the probability that the first is an even number and the second is 7? Fill in Joanne's solution to this question.

3	4	5	6
7	8	9	10

   At first, there are four cards with an even number and eight cards in total to choose from.

   So P(even) = _____.

   After one card with an even number has been drawn, there are seven cards left, and one of them is 7.

   So P(7) is _____.

   Then we multiply the two probabilities to get the probability of both events:

   P(even, 7) =

7. You choose a card randomly from this set of cards. Then you choose another card (without replacing the first). Find the probabilities.

   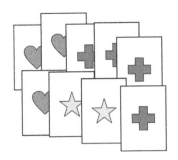

   a. P(heart, heart)

   b. P(star, cross)

   c. P(not heart, not heart)

   d. P(star, not star)

8. You choose two marbles randomly from the bag without replacing them.

   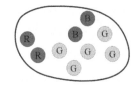

   a. What is the probability that both are green?

   b. What is the probability that neither is green?

   c. What is the probability that exactly one of the marbles is green?
   Hint: Do you want either of the events in (a) or (b) to occur?

244

9. Michael has 10 white socks and 14 black socks mixed together in a drawer. He randomly chooses one sock to wear and doesn't put it back. Then he chooses another sock. Find the probabilities:

   a. P(white, white)

   b. P(black, black)

   c. P(black, white)

   d. P(white, black)

   CHECK. The four probabilities above should total 1 or 100%.

   e. Add the probabilities in (a) and (b) to find the probability that Michael wears matching socks.

   f. What is the probability Michael *doesn't* wear matching socks?

10. You choose three cards from a standard 52-card deck (don't include the jokers). Calculate the following probabilities to the nearest tenth of a percent:

    a. What is the probability that all of them are aces (AAA)?

    b. What is the probability that the first of the three cards is an ace and the others are not (AXX)?

    c. What is the probability that the first of the three cards is not an ace, the second is, and the third is not (XAX)?

    d. What is the probability for XXA?

    e. Add the three probabilities from (b) through (d) to get the probability that exactly one of the three cards is an ace.

Puzzle Corner

Matthew has 8 white socks, 9 brown socks, and 10 black socks mixed together in a drawer. He chooses two socks randomly. Find the probability, to the nearest hundredth of a percent, that he gets to wear a matching pair.

Chapter 10 Mixed Review

1. Beth got a really unreasonable answer. Find what went wrong with her solution and correct it.

> Eighty liters of blueberries costs $35.
> How much would 52 liters cost?
>
> **Beth's Answer:** 52 liters would cost $118.86.
>
> **Beth's Solution:** $\dfrac{80}{35} = \dfrac{C}{52}$
>
> $35C = 80 \cdot 52$
>
> $35C = 4160$
>
> $\dfrac{35C}{35} = \dfrac{4160}{35}$
>
> $C = 118.86$

2. Write an equation using a variable for the unknown angle, and solve it.

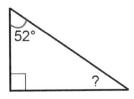

3. Determine whether the lengths 11.4 cm, 19 cm, and 15.2 cm form a right triangle.

4. Calculate the volume of this box both in cubic centimeters and in cubic meters.

$V = $ _____ cm³

$V = $ _____ m³

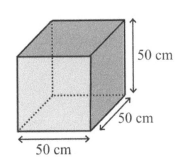

5. **a.** What kind of triangle is this?

   b. Find its area to the nearest ten square centimeters.

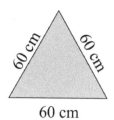

6. Margie drew this circle on a sheet of paper that measures 21 cm by 29.7 cm.

   a. Find the area of the circle to the nearest square centimeter.

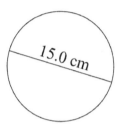

   b. What percentage of the area of the paper does the circle occupy? Give your answer to the nearest tenth of a percent.

   Note: don't use the rounded result from part (a) for this question as that may throw off your answer. For this calculation, you need to keep several decimal digits for the area of the circle.

7. A rectangle with sides of 3 1/2 in. and 2 in. is enlarged in a ratio of 2:5.

   a. Find the lengths of the sides of the resulting larger rectangle.

   b. Find the area of the resulting rectangle.

8. Use the distributive property to multiply.

a. $7(x + 8) =$	**b.** $4(2y - 10) =$	**c.** $0.1(2x + 18) =$

9. Sketch a rectangle with an area of $5x + 15$ and label the lengths of its sides.

10. Solve.

a. $\dfrac{v - 6}{7} = -31$

b. $\dfrac{x}{4} - 1 = -5$

11. In May, a bookstore sold 2,400 books. In June it sold 2,000 books. By what percentage did the book sales decrease?

12. The students in Seventh Grade took part in a math contest. Henry scored 77 points, and Mary got 86 points. The average score of all the students in the contest was 66 points.

   a. How much better (in percent) was Henry's score compared to the average score?
   Hint: Use relative difference.

   b. How much better (in percent) was Mary's score compared to the average score?

   c. How much better (in percent) did Mary score than Henry?

13. Solve. If the result is a fraction, simplify it to lowest terms.

a. $3 \cdot (-3) \cdot (-1) =$	b. $(-6) \cdot 8 \div 16 =$	c. $-8 \cdot (-2) \cdot (-5) =$
d. $(-42) \div (-7) =$	e. $7 \div (-42) =$	f. $-8 \div (-2) + (-5) =$

Chapter 10 Review

1. The chart lists the favorite school subjects of the students in a 7th grade classroom.

    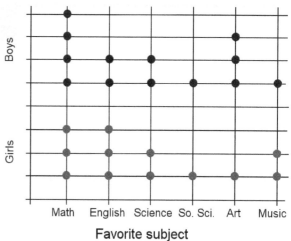

    a. You choose one student randomly. What is the probability that the student's favorite subject is not math, English, or science?

    b. What is the probability that a randomly chosen student's favorite subject is math?

    c. Now look at the boys only. If you choose one boy randomly, what is the probability that his favorite subject is math?

    d. If you choose a girl randomly from among the girls, then what is the probability that her favorite subject is math?

2. Abigail rolls two dice. Find the probabilities of these events:

    a. P(5, 6)

    b. P(even, even)

    c. P(at least 5, at least 5)

    d. P(at most 3, at most 2)

3. Julie and Jane experimented with rolling a die. They rolled the die 60 times in a row and recorded the results:

    2 4 3 3 5   2 5 4 1 6   1 4 4 4 5   4 2 6 4 1   5 2 1 1 1   2 1 3 6 2
    4 2 2 3 3   5 2 3 6 4   4 4 1 3 2   6 6 4 5 6   4 6 5 6 5   6 5 6 5 3

    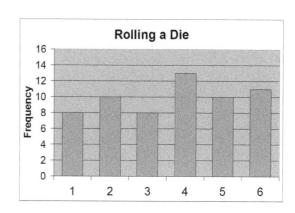

    a. In this experiment, what was the probability of rolling a 1?

    Rolling a 4?

    b. Why are those probabilities not 1/6?

249

4. Andrew did an experiment where he tossed two coins 200 times and recorded the outcomes. The table below shows his results. "H" means "heads," and "T" means "tails," so "HH," for example, means both coins landed "heads."

   a. Calculate the experimental and theoretical probabilities and fill in the table.

Outcome	Frequency	Experimental Probability (%)	Theoretical Probability (%)
HH	38		
HT	53		
TH	46		
TT	63		
TOTALS	200		

   b. How would the experimental probabilities change if Andrew redid this experiment with 2,000 tosses?

5. In a multi-player computer game, a computer chooses colors randomly for a girl's dress. The first color it chooses is the main color of the dress. The second color is for the bows and some layers of the skirt. The computer uses this list of colors: *red, blue, purple, pink, orange, yellow, mint*.

   After choosing the main color, the computer removes it from the list and chooses the second color from the resulting list of six colors. That way the dress is sure to have two different colors.

   a. What is the probability the computer chooses first purple, then orange?

   b. What is the probability the computer chooses first red, then not pink?

   c. Janet doesn't like mint. What is the probability her character gets a dress with no mint in it when she plays the game?

Chapter 11: Statistics
Introduction

Math Mammoth Grade 7 ends with a study of statistics. The chapter begins with a review lesson to remind students how to make a boxplot and a histogram and how to calculate the mean absolute deviation—all topics that were studied in 6th grade.

The first focus area of the chapter is random sampling. Students learn that sampling methods vary and that random sampling is likely to produce an *unbiased sample*—a sample that represents the population well. In the lesson *Using Random Sampling*, students choose several random samples from a population of 100 geometric shapes, and they see first hand that random samples can vary—even a lot. Yet if the sample size is sufficiently large, or if we have several random samples, we can be relatively confident in concluding something about the population itself. Students will also practice making inferences about populations based on several random samples.

The second major topic is comparing two populations, either directly or by using samples from the populations. Students learn to use the overall distributions and the measures of center and variability to compare two sets of data in various ways. While some of the ways in which we compare the data are informal only, all of the concepts presented are fundamental to the use of statistics in various sciences. Students also do a project where they gather data on their own from two populations and compare them.

If you are using the curriculum as a traditional pre-algebra program (as a preparation for algebra studies), you can consider omitting this chapter.

The Lessons in Chapter 11

	page	span
Review of Data Analysis	255	*5 pages*
Random Sampling	260	*4 pages*
Using Random Sampling	264	*7 pages*
Comparing Two Populations	271	*8 pages*
Comparing Two Samples	279	*6 pages*
Chapter 11 Mixed Review	285	*4 pages*
Chapter 11 Review	289	*3 pages*

Helpful Resources on the Internet

GRAPHS AND PLOTS

Statistics videos by Maria
Videos on statistics topics that are helpful for the lessons in this chapter.
http://www.mathmammoth.com/videos/statistics/statistics_lessons.php

Statistics Interactive Activities from Shodor
A set of interactive tools for exploring and creating boxplots, histograms, dot plots, and stem-leaf plots. You can enter your own data or explore the examples.

http://www.shodor.org/interactivate/activities/BoxPlot/

http://www.shodor.org/interactivate/activities/Histogram/

http://www.shodor.org/interactivate/activities/PlopIt/

http://www.shodor.org/interactivate/activities/StemAndLeafPlotter/

Analyzing and Displaying Data Gizmos from ExploreLearning
Gizmos are interactive online tools with lesson plans that allow you to explore and learn about the topic in a virtual, dynamic environment. This page includes gizmos for box-and-whisker plots, histograms, stem-and-leaf plots, polling, and more. The gizmos work for 5 minutes for free. You can also sign up for a free trial account.
https://www.explorelearning.com/index.cfm?method=cResource.dspResourceExplorer&browse=Math/Grade+6-8/Data+Analysis+and+Probability

Box-and-Whisker Plots Quiz
Review box-and-whisker plots with this interactive self-check quiz.
http://www.phschool.com/webcodes10/index.cfm?wcprefix=bja&wcsuffix=1203

Create A Box and Whisker Chart
An online tool for creating a box-and-whisker plot from your own data. Includes lots of options.
https://www.meta-chart.com/box-and-whisker

Stem-and-Leaf Plots Quiz
An online multiple-choice quiz that is created randomly. Refresh the page (or press F5) to get another quiz.
http://www.phschool.com/webcodes10/index.cfm?wcprefix=asa&wcsuffix=0905&area=view

Make Your Own Stem-and-Leaf Plot
Enter values from your own data, and this web page creates your stem-and-leaf plot for you.
http://www.mrnussbaum.com/graph/sl.htm

Comparing Data Displays
Practice interpreting and comparing dot plots, histograms, and box plots in this interactive online activity.
https://www.khanacademy.org/math/pre-algebra/pre-algebra-math-reasoning/pre-algebra-frequency-dot-plot/e/comparing-data-displays

Double Box-and-Whisker Plots Practice
Interactive practice questions about double box-and-whisker plots. The site may require a login with your Google or Facebook account or a free registration.
http://www.ck12.org/statistics/Double-Box-and-Whisker-Plots/asmtpractice/Double-Box-and-Whisker-Plots-Practice/

Displaying Univariate Data Practice
Interactive practice questions about distributions and various types of plots.
http://www.ck12.org/statistics/Displaying-Univariate-Data/asmtpractice/Displaying-Univariate-Data-Practice/

STATISTICAL MEASURES

Mean, Median, Mode, and Range
A lesson that explains how to calculate the mean, median, and mode for a set of data given in different ways. It also has interactive exercises.
http://www.cimt.org.uk/projects/mepres/book8/bk8i5/bk8_5i2.htm

Measures of Center and Quartiles Quiz from ThatQuiz.org
An online quiz about the measures of center and quartiles in boxplots, stem-and-leaf plots, and dot plots.
http://www.thatquiz.org/tq-5/?-jr0t0-l1-p0

GCSE Bitesize Mean, Mode and Median
Lessons with simple explanations and examples.
http://www.bbc.co.uk/schools/gcsebitesize/maths/statistics/measuresofaveragerev1.shtml

Quiz: Boxplots and Stem-and-leaf Plots
Practice interpreting graphs with this interactive online quiz.
https://www.thatquiz.org/tq/practicetest?pw4bt4kw18id6

Mean Deviation
A simple explanation about what the mean absolute deviation is, how to find it, and what it means.
http://www.mathsisfun.com/data/mean-deviation.html

Mean Absolute Deviation
Several videos explaining how to calculate the mean absolute deviation of a data set.
http://www.onlinemathlearning.com/measures-variability-7sp3.html

Working with the Mean Absolute Deviation (MAD)
A tutorial and questions where you are asked to create line plots with a specified mean absolute deviation.
http://www.learner.org/courses/learningmath/data/session5/part_e/working.html

SAMPLING

Random and Biased Sampling
A comprehensive lesson to read that explains about unbiased types of sampling.
(If the link does not work, then copy and paste it into your browser.)
http://www.ck12.org/na/Random-and-Biased-Sampling-in-a-Population---7.SP.1,2-1/lesson/user%3Ac2ZveDJAb3N3ZWdvLm9yZw../Random-and-Biased-Sampling-in-a-Population---7.SP.1%252C2/

Identify a Random Sample
What a sample and random sample. On the left of the web page are 3 other videos on (unbiased) and biased samples.
https://learnzillion.com/lesson_plans/5266-identify-a-random-sample

Analyze Numbers of Jubes
Test a machine that randomly packages three types of fruit jubes. Look at patterns in numbers of jube types.
https://fuse.education.vic.gov.au/Resource/LandingPage?ObjectId=c8dad524-5a86-4cf3-8ff4-cfe81813e230

Random or not? Analyse Runs of Jubes
Look at patterns in sequences of jube types, analyse the results of large samples, and much more.
https://fuse.education.vic.gov.au/Resource/LandingPage?ObjectId=2a4c7dd8-2319-46b2-a4d9-797174223fec

What Does Random Look Like?
Generate several sequences of twenty coin flips. Get a feel for what you would expect a random sequence to have.
http://nrich.maths.org/7250

Making Inferences about a Population by Analyzing Random Samples
This is a short instructional video about random sampling.
https://learnzillion.com/lesson_plans/6910-make-inferences-about-a-population-by-analyzing-random-samples

Random Sampling
Multiple-choice questions that test your understanding of the basics of random sampling.
https://www.khanacademy.org/math/recreational-math/math-warmup/random-sample-warmup/e/random-sample-warmup

Valid Claims
Practice figuring if we took a random sample and are we able to draw valid conclusions from the data.
https://www.khanacademy.org/math/probability/statistical-studies/statistical-questions/e/valid-claims

Making Inferences from Random Samples
Questions about what can reasonably be inferred, from a random sample, about an entire population
https://www.khanacademy.org/quetzalcoatl/content-improvement/middle-school-content/e/making-inferences-from-random-samples

Population and Sample (Random Sampling)
A detailed explanation about obtaining a random sample from a population. Includes practice problems.
https://www.learner.org/courses/learningmath/data/session1/part_d/index.html

Matchbox Machine: Take a Sample
take a sample of 100 matchboxes and make a boxplot to analyze the results, and much more.
https://fuse.education.vic.gov.au/Resource/LandingPage?ObjectId=38397b5d-25b8-4523-a0cd-0fc627a8a241

Fix the Matchbox Machine: Scoop Size and Speed
Check whether a machine is packing most matchboxes with an acceptable number of matches, look at boxplots made after taking samples of 100 matchboxes, and much more.
https://fuse.education.vic.gov.au/Resource/LandingPage?ObjectId=4087511a-60a5-48ec-b866-998882cc83cb

Polling: Neighborhood Gizmo
Conduct a phone poll of citizens in a small neighborhood to determine their response to a yes-or-no question. The gizmo works for 5 minutes for free. A free trial account is also available.
http://www.explorelearning.com/index.cfm?method=cResource.dspDetail&ResourceID=507

Make inferences about a population by analyzing random samples
A video lesson that teaches how to make inferences about a population based on random samples.
https://learnzillion.com/lesson_plans/6910-make-inferences-about-a-population-by-analyzing-random-samples

Capture-Recapture Method
Learn how this method for estimating population size works. The second link is an interactive simulation.
https://nrich.maths.org/9609

http://www.cengage.com/biology/discipline_content/animations/capture_recapture.html

COMPARING TWO POPULATIONS

Comparing Populations - Khan Academy
Multiple-choice questions to practice comparing centers of distributions in terms of their spread.
https://www.khanacademy.org/math/cc-seventh-grade-math/cc-7th-probability-statistics/cc-7th-population-sampling/e/comparing-populations

Grade 7 Mathematics Module 5, Topic D, Lesson 22
Students express the difference in sample means as a multiple of a measure of variability, and learn what a difference in sample means can show about the population means.
https://www.engageny.org/resource/grade-7-mathematics-module-5-topic-d-lesson-22

Grade 7 Mathematics Module 5, Topic D, Lesson 23
Students use data from random samples to draw informal inferences about the difference in population means.
https://www.engageny.org/resource/grade-7-mathematics-module-5-topic-d-lesson-23

FACTS AND FIGURES

GapMinder
Visualizing human development trends (poverty, health, gaps, income) This is an interactive, dynamic tool.
https://www.gapminder.org/tools

WorldOdometers
World statistics updated in real time. Useful for general educational purposes—for some stunning facts.
http://www.worldometers.info

Review of Data Analysis

In this lesson we review some statistics topics from earlier grades.

A **boxplot** (also called a box-and-whisker plot) is a handy way to show graphically how the data is spread and where its center (median) is. It can be drawn horizontally or vertically. To draw a boxplot, we use the so-called **five-number summary**, which consists of the minimum of the data plus the four quartiles. The four quartiles divide the data into quarters:

- The first (or lower) quartile is the median of the lower half of the data.
- The second quartile is the median.
- The third (or upper) quartile is the median of the upper half of the data.
- The fourth quartile is the maximum of the data.

The boxplot on the right is made from the Jones family's monthly expenditures over a 12-month period. For example, we can see that their maximum monthly expenditure was a little under $1,250.

The **interquartile range** is the difference between the first and third quartiles. Since the quartiles divide the data into quarters, the middle half of the data (from 25% to 75%) lies within the interquartile range (the "box"). The interquartile range is a measure of **variability**: the larger the interquartile range, the more variable (spread out) the data is.

In the case of Jones family's expenditures, the interquartile range is $1195 − $1105 = $90. Since this number is small compared to the median of $1,105, it indicates that the data is spread out very little. Most of the time, the Jones family's expenditures do not vary a lot from month to month.

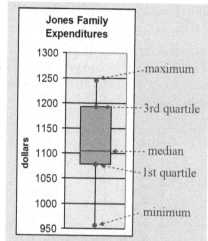

Five-number summary:

Minimum: $956
First quartile: $1,077.50
Median: $1,105
Third quartile: $1,192.50
Maximum: $1,245

Example 1. Let's make a boxplot of the ages of the children in a softball club. Their ages are:
4, 4, 5, 5, 5, 6, 6, 6, 7, 7, 7, 7, 7, 8, 8, 8, 8, 8, 9, 9, 11.

Here is the data divided into upper and lower halves. Since there is an odd number of children, we will leave the median age (7) by itself, so it belongs to neither the lower nor the upper half.

$$4, 4, 5, 5, 5, 6, 6, 6, 7, 7, \quad 7, \quad 7, 7, 8, 8, 8, 8, 8, 9, 9, 11$$

The first quartile is the median of the lower half of the data, which is the average of 5 and 6, or 5.5 years old. Similarly, the third quartile is the median of the upper half of the data, 8 years old. The interquartile range is the difference between those two: 8 years old − 5.5 years old = 2.5 years.

Notice that we include the unit (in this case "years old" or "years") with all of these quantities.

Now we can draw the boxplot:

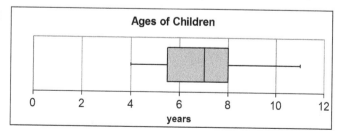

Overall, the data has a medium spread: the interquartile range of 2.5 years is neither especially large nor especially small compared to the median age of 7 years old. Since the second half of the box is fairly short, the data is concentrated there: there are a lot of 7- and 8-year olds.

Histograms are like bar graphs, but the bars are drawn touching each other. They are used with numerical data.

Example 2. The data below lists the prices, in dollars, of hair dryers in three different stores. Make a histogram.

14 15 19 20 20 20 21 24 25 34 34 35 35 37 42 45 55

First we need to decide how many bars to make. For that, we calculate the **range**, or the difference between the maximum and the minimum: $55 - $14 = $41. Then we divide the range into equal parts called **bins**.

If we make five bins, we get $41 ÷ 5 = $8.2, which means the bins would be 8.2 dollars apart. However, in this case it is nice to have the bins go by whole numbers, so we round $8.2 up to $9 and use $9 for the bin width. Each successive bin starts at $9 more than the previous bin, and there won't be any gaps between the bins.

The important part is that all the data values need to be in one of the bins. You may have to try out slightly different bin widths and starting points to see what works best. In this case, starting the first bin at $13 makes the last bin end at $57, which works, because the data will all "fit" into the bins. (Starting at $14 would work, too.)

Price ($)	Frequency
13..21	7
22..30	2
31..39	5
40..48	2
49..57	1

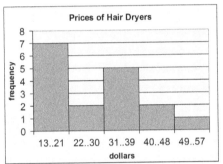

Mean absolute deviation (MAD for short) is a measure of variability. It tells us how much the individual data items differ from their mean, on average. The term "mean absolute deviation" actually explains itself:

- *Deviation* is a difference: we look at how much the individual data values deviate (differ) from the mean.

- *Absolute* deviation means the absolute value of each deviation. We take all of the deviations (differences) as positive.

- *Mean* absolute deviation is the mean (average) of all the absolute deviations.

Example 3. Calculate the mean absolute deviation for this data set: 5, 6, 6, 6, 6, 7, 7, 8, 8, 9.

First we calculate the mean itself:
Mean = (5 + 6 + 6 + 6 + 6 + 7 + 7 + 8 + 8 + 9)/10 = 6.8.

Next, we calculate how much each data value deviates from the mean of 6.8. The table on the right lists the absolute deviations.

Lastly, we calculate the mean of the absolute deviations:
MAD = (1.8 + 4 · 0.8 + 2 · 0.2 + 2 · 1.2 + 2.2)/10 = 1.0

So the mean absolute deviation is 1.0. This means that, on average, the individual data values differ from the mean of 6.8 by 1. This value (1.0) is not especially large compared to the mean (6.8), so the data is fairly concentrated (not spread out). We can even see that from the list of data values.

Value	Absolute deviation from the mean
5	1.8
6	0.8
6	0.8
6	0.8
6	0.8
7	0.2
7	0.2
8	1.2
8	1.2
9	2.2
mean 6.8	**MAD 1.0**

It is much quicker to calculate the MAD using a spreadsheet program. For your reference, here are the instructions for how to do it in Excel.

1.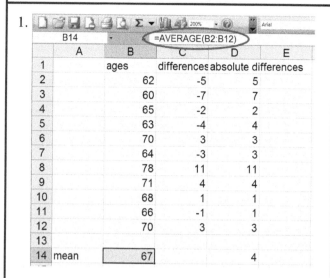

To calculate the mean of a set of data, in the cell where you want the calculation to appear, type:

=AVERAGE(B2:B12)

When you type the formula in the cell, it appears in the formula bar at the top, as in the image. A formula always starts with an equals sign.

Press "**ENTER**" to see the answer, 67.

2.

Next we calculate the difference between each item of data and the mean.

Type "=B2 - B14" to subtract the values in cells B2 and B14.

The dollar signs in **B14** make it an **absolute reference**, so it doesn't change when you copy and paste the formula into other cells. Pasting the formula into the cells below is a quick way to get the spreadsheet to calculate those values, too.

3.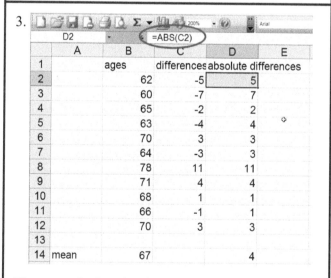

Now we calculate the absolute value of each difference.

In cell D2 type "=ABS(C2)" to calculate the absolute value of the number in cell C2. When you paste cell D2 into the cells below it, Excel automatically changes the reference to cell C2 to the correct cell.

4.

Lastly, we are ready to calculate the mean absolute deviation by taking the average of the values in cells D2 to D12. In the cell where you want the value to appear, type "=AVERAGE(D2:D12)".

The answer "4" will then appear in the cell after you press "**ENTER**."

1. **a.** Read the five-number summary from the boxplot, and give the interquartile range.

   Minimum:
   First quartile:
   Median:
   Third quartile:
   Maximum:
   Interquartile range:

   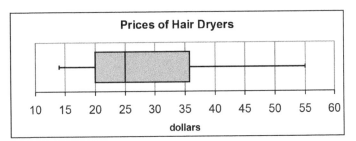

   b. Describe the overall variability (spread) of this data.

2. Calculate the five-number summary and draw a boxplot for the following set of data.

   The ages of participants in a children's fortune wheel game:
   6, 8, 9, 10, 10, 10, 11, 11, 11, 11, 12, 12, 12, 12, 13, 13, 13, 15, 18, 22

Five-number summary
Minimum _____
First quartile _____
Median _____
Third quartile _____
Maximum _____
Interquartile range _____

3. The table lists the highest daily temperatures for January.

Mon	Tue	Wed	Thu	Fri	Sat	Sun
−2°C	−4°C	2°C	3°C	1°C	−6°C	−5°C
0°C	−3°C	−3°C	−2°C	−10°C	−12°C	−13°C
−1°C	1°C	3°C	2°C	0°C	−2°C	−8°C
−12°C	−11°C	−9°C	−10°C	−5°C	−3°C	−4°C
−2°C	−2°C					

a. Make a histogram of the data. Use five bins.

b. Calculate the mean to one decimal digit.

c. Lastly, calculate the mean absolute deviation (MAD) to two decimal digits.

Random Sampling

When researchers have a question concerning a large population, they obtain a **sample** (a part) of that population. That is because it is typically impossible to study the entire population.

For example, if you want to know how the citizens of France feel about climate change, you cannot just go and ask every person in France about it. You would choose for example 600 French citizens as your sample and ask them your question.

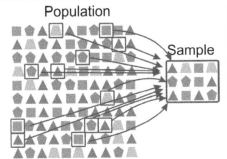

The way a sample is chosen is very important. Some methods of sampling may produce a sample that is *not* representative of the entire population. We call that a **biased sample**.

For example, if you are studying a student population of 630 in a school with close to an equal number of boys and girls, and you happen to choose a sample of 20 boys, then your sample is biased. It doesn't represent the entire population well.

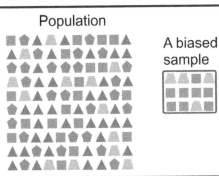

We need to use **unbiased sampling methods** in order to get a sample that truly represents the population being studied. The best way to avoid biased samples is to select a **random sample**.

The main characteristics of a random sample are:

1. **Randomness:** each member of the population has an equal chance of being selected.

    Let's say a researcher is studying the types of cars Americans own. He decides to interview only people he finds at a local mall because that mall is close to where he lives, so it is convenient for him. His sample is biased because not every member of the US population even has a chance to be selected in his sample. Maybe the people at his local mall are predominantly rich people who own several cars per family, so in that respect those people would not be a good representation of the entire population of the US.

    We call this type of sample a **convenience sample** because it is convenient or easy to obtain.

2. **External selection:** respondents must be chosen by the researcher, not self-selected.

    If our researcher mails a questionnaire to various people across the US asking them to fill it out and return it, his sample is a **voluntary response sample**, which is a biased sample. Some people volunteer to return the questionnaire, but others don't. The people themselves decide whether or not to be a part of the sample.

    Why might this be a problem? Some of the people who would choose to take part may have an external reason to do so. They might want to show off how "good" they are in the particular aspect being studied, or they might just like to speak out about their opinions.

Our researcher could get a true random sample by choosing people randomly from a list of people living in the US and calling them. That way, each person has an equal chance of being selected in the sample (it is random), and the people cannot self-select to take part (the researcher chooses who takes part).

An unbiased sampling method is more likely to produce a representative sample.

1. You are studying whether students in a large college prefer to drink coffee black, with milk, with cream, or with sweetener, or whether they prefer not to drink coffee at all.

   a. Which of the six sampling methods listed below produce a voluntary response sample?

   b. Which methods don't give each member of the student population an equal chance to be selected for the sample?

   c. Which method is likely to produce a sample with only coffee drinkers, overlooking those who don't drink coffee?

   d. Which method will be the most likely to give you a representative (unbiased) sample?

   Sampling Methods

   (1) You interview 80 students in a cafe on the campus.

   (2) You interview 80 students who come in at the main door of the campus.

   (3) You interview the first 80 students you happen to meet on a certain day.

   (4) You choose 80 names randomly from a list of all the students. You call them to interview them.

   (5) You send an email to all the students in the college, asking them to fill in a form on a web page you have set up. You hope to get at least 80 responses.

   (6) You choose 80 names randomly from a list of all the students. You send them an email, asking them to fill in a form on a web page you have set up.

2. A recipe website posts a poll on their home page that any visitor to that website can take. In it, they ask if people are looking for a recipe for a dessert, a main dish, a side dish, bread, or salad. During the course of one Sunday, 4,600 people visit the page, and 252 of them fill in the poll. Explain why the poll results will be based on a biased sample.

> Some common random sampling methods are:
>
> 1. **Simple random sampling.** Each individual in the sample is chosen randomly and entirely by chance, perhaps by using dice, through pulling names out of a hat, or with a random number generator.
>
> 2. **Systematic random sampling.** The individuals of the population are placed in some order, and then each individual at a certain specified interval is selected for the sample.
>
>   For example, a supermarket might study the shopping habits of its customers by choosing every 15th customer who enters the store for the sample.
>
> 3. **Stratified random sampling.** The population is first divided into categories (strata) and then a random sample is obtained from each category.
>
>   For example, to study how much sleep students in a particular school get, you might first divide the students into groups by grade levels (the stratification), then select a random sample from each of the grade levels.

3. A population to be studied doesn't have to be of people. A factory produces MP3 players. Out of the 500 units that the factory produces each day, a quality control inspector selects 25 for testing to study their quality and reliability. Which way should he choose those 25 so that his sample would best represent all the MP3 players that the factory produces?

   a. Choose the first 25 produced on a given day.

   b. First choose a number between 1 and 20 randomly. Select the player corresponding to that number, and after that, every 20th player, in the order they were produced that day.

   c. Choose 25 players that have just been finished around 1 PM when the inspector is touring the factory.

   d. Generate 25 random numbers between 1 and 500 and choose the corresponding 25 MP3 players in the order they were produced that day.

4. Ryan has two large fields planted with green beans. He wants to compare the bean plants in one field with the plants in the other. Design a practical sampling method for him to produce an unbiased sample.

> **Sometimes it is not obvious how a particular sampling method might be biased.**
>
> If you are studying students' homework habits in a particular school, it might initially make sense to interview the first 25 students who come into the school in the morning. However, there could be an underlying factor that makes that method biased. What if students who are diligent with their homework also tend to come to school early? In that case, students who are not diligent don't have an equal chance of being selected for your sample. A better method is to use systematic random sampling and to choose, say, every 10th student entering the school for the sample.

5. Heather is studying the effect of how the method of feeding affects the health of a baby during its first year of life. She has already determined that babies who are fed with infant formula get sick more often than babies who are fed with human milk, but she especially wants to find out how often babies who are fed both formula and breast milk get sick. Explain why interviewing mothers in the places below will produce a biased sample:

   a. A pediatrician's office.

   b. A breastfeeding class for new moms.

6. **a.** Choose a sampling method that will produce a biased sample based on self-selection and explain how that would happen, based on Heather's situation in Question 5.

   b. Design a sampling method for Heather that is most likely to produce a representative sample.

7. An organization that helps teenagers with drug problems has set up a telephone hot line for teens to call in to discuss their problems. After a few months of operations, the organization wants to evaluate the effectiveness of their service. Since they don't usually get as many calls on Tuesdays, they decide to choose a particular Tuesday to ask each teen at the end of the call to answer a few questions about how the service has helped. Is this a good method for selecting a sample? Explain.

Using Random Sampling

1. In this activity, you will make several samples of 10 from this population of shapes:

   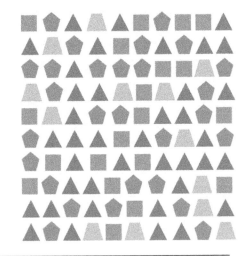

   Since the shapes are in a 10 by 10 grid arrangement, you can easily assign a number from 1 to 100 for each shape. To obtain a random sample, you can use one of these ideas or come up with your own.

   - Choose a random number between 1 and 10. Then, starting from that number, choose every 10th shape.

   - Go to **https://www.random.org/integers/** and generate 10 random integers between 1 and 100. (If the set of numbers contains a duplicate, discard that set and make another.)

   Here is an example sample (Sample 1) to help you get started:

   It is based on generating these random numbers at the website above: 76 17 51 63 88 29 95 73 40 69

   Generate at least five more samples. Count the number of each kind of shape in each of your samples, and fill in the table. Lastly, calculate the average number of triangles, the average number of squares, and so on.

	Sample 1	Sample 2	Sample 3	Sample 4	Sample 5	Sample 6	Averages
triangles	5						
squares	1						
pentagons	3						
trapezoids	1						
Total	10	10	10	10	10	10	—

2. Now imagine that you haven't seen the entire population of shapes but you try to infer (conclude) something about the entire population of shapes based on these six samples.

   a. Which shape seems to be the most common?

   b. Which shape seems to be the least common?

   c. List the shapes in order, from the least common to the most common:

   d. Based on the average number of the shapes in the six samples, *estimate* how many of each shape there are in the entire population of 100 shapes:

   _____ triangles   _____ squares   _____ pentagons   _____ trapezoids

Here are some important points to realize and remember concerning random sampling. You probably noticed these facts while doing the activity:

1. **Random samples vary.** The differences occur simply because their members are indeed chosen from the population at random.

2. **A random sample is more likely to be unbiased, but that is not 100% certain.** In other words, it is possible for a random sample *not* to be representative of the population as a whole. However, the chances of a random sample being unbiased are greater than the chances of a non-random sample being unbiased.

3. **It is better to base inferences about the population on more than one sample.** However, if you cannot obtain more than one sample, then it is recommended to increase the sample size as much as possible, because a large sample represents the whole population better than a small one.

Example 1. Three people are running for mayor in a town with 20,000 voters. Two companies conducted separate polls of 350 people, asking who they would vote for in the final election. Here are the results:

	Smith	Harrison	Jones
Poll 1	63	220	67
Poll 2	53	238	59

Based on these results, what can we conclude about the results of the final election?

In both polls, Harrison is winning and by a large margin. In other words, far more people are claiming that they will vote for Harrison than for Smith or Jones. So we can fairly confidently conclude that Harrison will be the winner of the actual election.

Not only that, in both polls, Jones did better than Smith. So it is likely, but not sure, that Jones will beat Smith in the actual election. We cannot say that for sure because the differences are small: 4 votes and 6 votes.

We can also quantify the election results (use actual numbers). In Poll 1, 220 is more than three times 63 or 67. In Poll 2, 238 is more than four times 53 and about four times 59. Based on those, we can say that Harrison will get roughly 3-4 times more votes than either of his opponents.

Example 2. The data below presents the results of three different samples from a study about how students prefer to drink coffee.

	Black	With milk	With cream	Milk and sugar	Cream and sugar	Totals
Sample 1	12	21	24	36	37	130
Sample 2	9	23	22	37	39	130
Sample 3	14	18	20	36	42	130

What can we infer based on the data?

(1) Looking at the numbers carefully we can see that in each of the samples "Cream and sugar" was the winner and "Milk and sugar" came fairly close behind it.

(2) The two options "With milk" and "With cream" are also close to each other, but we cannot say for sure which of them is preferred, because in Samples 1 and 3, "With cream" beats "With milk", whereas in sample 2 it is the opposite way.

(3) Drinking black coffee is the least popular option in all three samples.

(4) We can quantify the results. For example, "Cream and sugar" is the favorite of roughly 40/130 = 4/13 of the students, and 4/13 is almost 4/12 = 1/3. So we can state that almost 1/3 of the students prefer to drink their coffee with cream and sugar. You can make similar statements using approximate fractions for the other options.

> **What kinds of inferences can you make about the entire population based on random samples?**
>
> Based on what the data demonstrates, you may be able to…
>
> - state which option is the most or least, the best or worst, the winner or loser, *etc.*
> - compare two options as better or worse, more or less, *etc.*
> - quantify the above statements with numbers, fractions, or percentages:
>   *How much* more or less is one option than another?
> - find trends: identify an increase or a decrease in some quantity as some other quantity, such as time, increases or decreases.

3. A large workplace conducted a survey of their employees' sleeping hours. They took two samples of 65 people, one week apart. What can you infer based on these results?

	< 5 h	5 h	6 h	7 h	8 h	> 8h	Totals
Sample 1	1	4	21	32	6	1	65
Sample 2	2	8	23	26	4	2	65

4. A music band wanted to find out which of their songs their audience likes best. They randomly chose some people to be interviewed after two of their concerts, asking them what their favorite song was. The results are in the table at the right.

   What conclusions can you draw from the data?

Songs	Concert 1 (Sample 1)	Concert 2 (Sample 2)
"Love You"	4	3
"My Best"	9	11
"Never Again"	7	5
"Sunshine"	5	6
Totals	25	25

Example 3. Let's go back to our example about the three candidates running for mayor. If 15,000 people end up voting for mayor in the actual election, estimate the number of votes each candidate will get.

First we calculate the percentage of votes each candidate got in each poll:

	Smith	Harrison	Jones	Total
Poll 1	63	220	67	350
Percentage	18.00%	62.86%	19.14%	100%
Poll 2	53	238	59	350
Percentage	15.14%	68.00%	16.86%	100%
Averages:	**16.57%**	**65.43%**	**18.00%**	**100%**

Smith: Based on Poll 1, we would estimate that he would get $0.18 \cdot 15{,}000 = 2{,}700$ votes. Based on Poll 2, our estimate would be $0.1514 \cdot 15{,}000 = 2{,}271$ votes. Let's use the average percentage of 16.57% based on both polls. We estimate he will get around $0.1657 \cdot 15{,}000 = 2{,}486$ votes $\approx$ <u>2,500 votes</u>.

Then we can use the estimates from the individual polls to gauge how far off our estimate of 2,500 votes is. These numbers (2,700 and 2,271 votes) differ from 2,500 votes by about 200-300 votes. Since the samples are random, the estimate of 2,500 votes may be off by 2-3 times the 200-300 votes; we cannot really know without having dozens of samples. So, we state that our estimate may be <u>off by several hundred votes</u>.

Harrison: Based on Poll 1, he would get $0.6286 \cdot 15{,}000 = 9{,}429$ votes. Based on Poll 2, he would get $0.68 \cdot 15{,}000 = 10{,}200$ votes. Using the average percentage, we estimate he will get around $0.6543 \cdot 15{,}000 = 9{,}815$ votes $\approx$ <u>9,800 votes</u>.

The estimates of 9,429 and 10,200 votes differ from the estimate of 9,800 votes by about 400 votes. Again, the real value may differ from the estimate several times the 400 votes. Based on that, we gauge that the estimate of 9,800 votes may be <u>off by over a thousand votes</u>.

Jones: Based on Poll 1, he would get $0.1914 \cdot 15{,}000 = 2{,}871$ votes. Based on Poll 2, he would get $0.1686 \cdot 15{,}000 = 2{,}529$ votes. Using the average percentage, we estimate he will get around $0.18 \cdot 15{,}000 =$ <u>2,700 votes</u>.

The two estimates of 2,871 votes and 2,529 votes differ from that by a few hundred votes. We might gauge that the estimate of 2,700 votes may be <u>off by hundreds of votes</u>.

5. **a.** Let's continue quantifying the results of the study about how students prefer to drink coffee. Use statements with fractions, such as "about 1/5" and "slightly more/less than 1/10."

	Black	With milk	With cream	Milk and sugar	Cream and sugar	Totals
Sample 1	12	21	24	36	37	130
Sample 2	9	23	22	37	39	130
Sample 3	14	18	20	36	42	130

_____ of the students prefer to drink coffee black.

_____ of the students prefer to drink coffee with milk but no sugar.

_____ of the students prefer to drink coffee with cream but no sugar.

_____ of the students prefer to drink coffee with milk and sugar.

5. **b.** Calculate the missing percentages to the hundredth of a percent.

	Black	With milk	With cream	Milk and sugar	Cream and sugar	Totals
Sample 1	12	21	24	36	37	130
Percentage						
Sample 2	9	23	22	37	39	130
Percentage						
Sample 3	14	20	18	35	43	130
Percentage						
Average %						

c. Estimate how many students in a group of 500 will prefer to drink their coffee black. Also, gauge how far off your estimate might be.

d. Estimate how many students in a group of 500 students will prefer to drink their coffee with milk (no sugar). Also, gauge how far off your estimate might be.

e. Estimate how many students in that group of 500 will prefer to drink their coffee with cream and sugar. Also, gauge how far off your estimate might be.

6. A certain factory has 2,150 employees. The factory surveyed their workers about their sleeping habits. They used two samples of 65 people each.

	< 5 h	5 h	6 h	7 h	8 h	9 h	Totals
Sample 1	1	4	21	32	6	1	65
Sample 2	2	8	23	26	4	2	65

   a. Estimate the number of employees in the factory who get 7 hours of sleep. Gauge how far off your estimate is.

   b. Estimate the number of employees who get less than 5 hours of sleep. Gauge how far off your estimate is.

7. Scientists studied driving speeds on a popular stretch of highway. Here is their data.

Speed (mph)	46-50	51-55	56-60	61-65	66-70	>71	Total
Sample 1	46	180	51	13	8	2	300
Sample 2	44	167	67	15	7	0	300

   a. What inferences can you make?

   b. Based on the data, out of 1000 drivers predict how many would be driving at a speed of 51 to 55 miles per hour. Also, gauge how far off your estimate might be.

8. Study the lengths of words in a science textbook. Take six random samples of 100 words. For example, you could randomly choose six pages and take the first 100 words at the beginning of each of those pages.

   a. Record your results in a table.

   b. Estimate the mean word length in that science textbook based on your samples.

   c. How far off might your estimate be?

Comparing Two Populations

In this lesson, we will use measures of center and variability to compare two sets of data.

Example 1. The two dot plots on the right show the ages of two different groups of children. The top plot shows a group of children of ages 5 to 7, and the bottom plot a group of ages 3 to 5.

The center of the top plot is its median at 6 years old. The center of the bottom plot is its median at 4 years old. Each distribution is clustered about its median with little variability (spread). Although both groups include 5-year-old children, there is otherwise no overlap in ages.

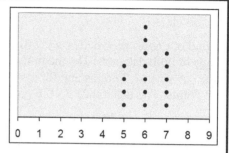

median 6 years   IQR 1.5 years

Intuitively we notice that the ages of these two groups are distinctly different. In statistical terms, we would say that there is a **significant difference** in the ages of the two groups.

There is a way to quantify the significance of the difference numerically: **compare the difference in the measures of center to the measure of variability**.

The measures of center—the medians—are 6 years old and 4 years old, so the difference between them is 2 years. The measure of variability—the interquartile range or IQR—of the first group is 1.5 years and of the second, 1 year. Let's round to 1 year in the comparison.

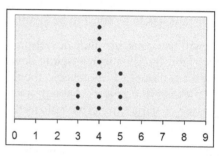

median 4 years   IQR 1 year

The difference in the medians (2 years) is about twice the measure of variability (about 1 year). This means that the difference is indeed significant. If the difference in centers had been only a fraction (1/2 or less) of the measure of variability, then the difference would not have been significant.

Example 2. The two dot plots on the right again show the ages of two different groups of children.

This time, we can see from the plots that there is great variability in the ages in both groups. The data is very spread out. The data sets also overlap a lot: the first group has children from 1 to 12 years old, and the second from 2 to 13 years old, which means the overlap is from 2 to 12 years old.

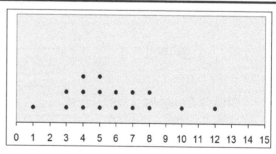

median 5 years   IQR 3.5 years

The difference in the medians is 2 years. However, the interquartile range is much bigger now (3.5 and 4.5 years; we can round it to 4 years). <u>Therefore, the difference in the medians (2 years) is only about 1/2 of the measure of variability</u>.

This fact, along with the large overlap, helps us to see that the difference in the medians is not very significant. The ages of these groups of children are not greatly different.

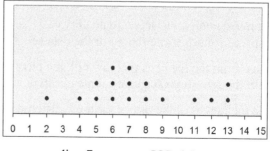

median 7 years   IQR 4.5 years

Example 3. The two graphs show science test scores for two classes, 7-A and 7-B. Which class did better, generally speaking?

We can see the answer just by looking at the distributions: The bars in the graph for 7-B are more skewed towards the right than the bars in the graph for 7-A. So class B did better.

To find out *how much* better, we will <u>compare the means</u> of both data sets. The mean test score for class 7-A was 64.2 points and for class 7-B it was 74.8 points. The difference is 10.6 points.

Is that a significant difference?

To check that, we compare the difference of 10.6 points to how variable or spread out the distributions are. The more variability there is in the two distributions, the bigger the difference in the means has to be for it to be significant.

We will use mean absolute deviation as a measure of variability. The mean absolute deviation of each data set is close to 11 points. So the difference in the two means (10.6 points) is <u>about one time the measure of variability</u>. That tells us that the difference *is* significant.

If the difference in the means had been, say, 0.3 times the measure of variability (only 3 points), then the difference wouldn't have been significant.

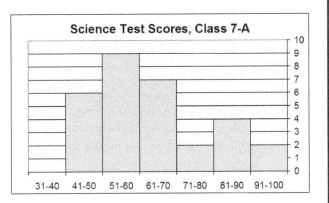

mean 64.2   MAD 11.8

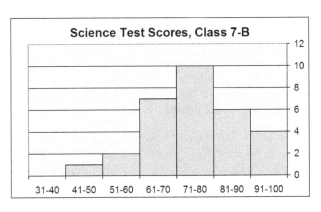

mean 74.8   MAD 10.3

Example 4. The boxplots show the prices of 1000-piece puzzles in two stores, ToyLand and Child's Delights.

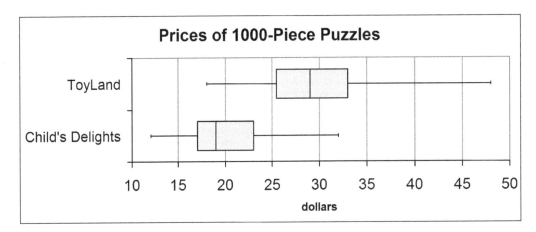

Boxplots make comparing sets of data very easy, since you can immediately see both the centers (medians) and the spread (interquartile range) of the data sets from the plot.

We can see from the medians that, overall, the puzzles in Child's Delights are cheaper. The prices in Toyland vary more, though, so you can find some cheap puzzles there, as well.

But while the ranges of the prices are quite different, the interquartile ranges (the lengths of the boxes) are similar: about 8 dollars for ToyLand and about 6 dollars for Child's Delights. The difference in the medians is about 10 dollars, which is about 1.5 times the interquartile range. That is a significant difference.

1. Jim studied the rain patterns in his home town. He made boxplots to show the number of days with rain in September and October from data collected over 20 years.

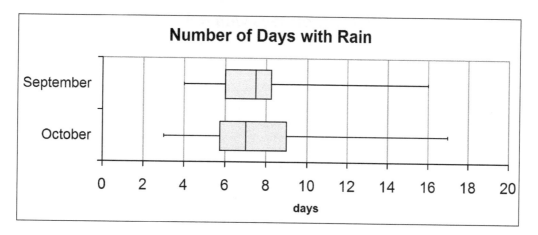

   a. Describe the overlap of the two distributions.

   b. Based on the medians, overall which month has more days with rain?

   Which month has the greater variability in the number of days with rain?

   c. Estimate from the plot the medians for October and September and their difference.

   median (September)_____   median (October)_____   difference _____

   d. Estimate from the plot the interquartile ranges for October and September.

   IQR (September) _____   IQR (October) _____

   e. Based on your answers to (c) and (d), is the difference in the medians significant?

2. Jim also made boxplots to compare the number of days with rain in March and September.

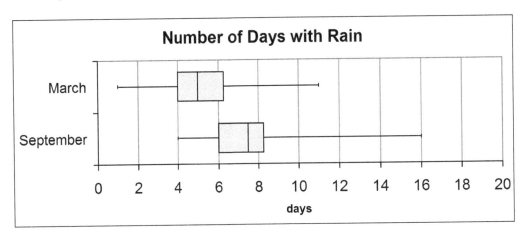

a. We can see that these two distributions overlap each other quite a bit, but not completely.
Let's say that a certain month had 4 rainy days.
Which is more likely, that the month was March or September?

b. Let's say that a certain month had 8 rainy days.
Which is more likely, that the month was March or September?

c. Based on the plots, overall which month has more days with rain?

Which month has the greater variability in the number of days with rain?

d. Find the difference in the medians and the interquartile ranges.

e. Compare the difference in the medians to the variability of the data.

The difference in the medians is about _____ times the interquartile range.

Are there significantly more days with rain in September than in March?

3. The following are the science grades of two 7th grade classes. This school grades on a five-point system where 5 = A, 4 = B, 3 = C, 2 = D, and 1 = F. Make bar graphs from the data.

Class 7-A	
Grades	Students
1	5
2	8
3	7
4	5
5	2

Class 7-B	
Grades	Students
1	3
2	6
3	7
4	7
5	4

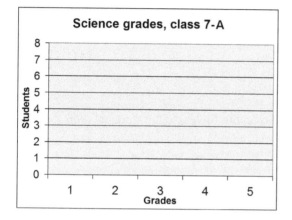

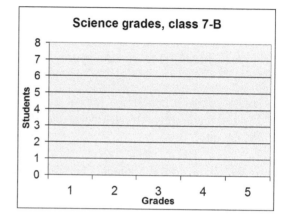

a. Compare the bar graphs visually.
Based on the graphs, did either class do better overall? If so, which one?

Did either class have more variability in the grades? If so, which one?

b. Now, calculate the mean of the grades for each class and the difference in the means.

Class A, mean: _____   Class B, mean: _____   Difference: _____

c. The mean absolute deviations of the data are 1.01 (class A) and 1.02 (class B).
The numbers are quite close. This means the variability is similar in both sets of data.

Compare the difference in the mean scores to the variability of the data, and use that to explain whether one of the classes did *significantly* better than the other.

275

4. Mrs. Ross gave her calculus class three quizzes. The bar graphs for the scores are below.

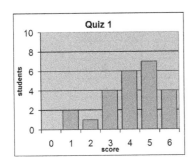 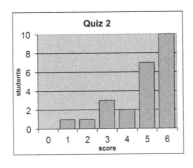 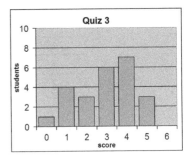

         mean _____   mean _____   mean _____

         MAD 1.14   MAD 1.13   MAD 1.22

a. Look at the graphs. Mrs. Ross felt one of the quizzes turned out too easy (the students didn't!). Which one?

b. In which quiz did the students fare the worst?

c. The mean scores for the three quizzes were: 2.96, 4.13, and 4.79.
Match each mean with the correct graph.

d. Compare quiz 2 and quiz 3 now. What is the difference in the means for quiz 2 and quiz 3? _____

This difference is about _____ times the mean absolute deviation of the data (1.13).

Is the difference in the means significant?

5. The data below give you the heights in inches, of the players in a women's field hockey team and in a women's basketball team.

   a. Determine the five-number summaries and draw side-by-side boxplots for the data.

Field Hockey Team Heights (inches)	
60	66
62	66
62	66
62	67
62	67
63	68
63	68
64	68
64	68
64	70
65	

   Five-number summary

   Minimum: _____

   1st quartile: _____

   Median: _____

   3rd quartile: _____

   Maximum: _____

   Interquartile range: _____

Basketball Team Heights (inches)	
67	72
68	72
69	72
69	74
69	75
70	76
71	

   Five-number summary

   Minimum: _____

   1st quartile: _____

   Median: _____

   3rd quartile: _____

   Maximum: _____

   Interquartile range: _____

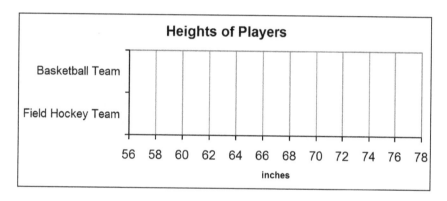

   b. Describe the overlap of the distributions.

   c. Let's say you met a female athlete who was 70 inches tall. Which is more likely, that she plays field hockey or that she plays basketball?

   d. Based on the boxplots, overall which team appears to be taller?

   Which team appears to have greater variability in the heights?

   e. Is the difference in the medians significant?

   Justify your reasoning.

6. Over the course of a month, a shoe store chain studied the sales of a "Wonder Shoe" by shoe size in two different stores. They did that for two purposes: (1) to check if there was any difference in the sales in the two stores, and (2) to obtain data to know how many of each size to order from their distributor.

   a. Underline the correct choice.
   There is (not much / some / a lot) of overlap in these distributions, and overall, they appear (different / similar).

   b. Calculate the difference in the mean shoe size sold in these two stores

   difference _____

   c. Compare the difference in the mean scores to the variability of the data, and use that to explain whether the difference in the mean scores is significant or not.

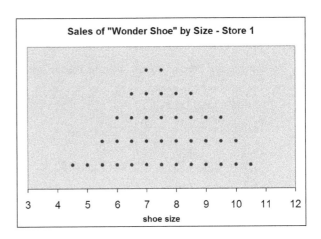

mean 7.61   MAD 1.20

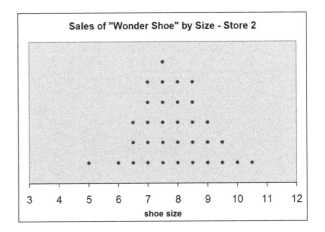

mean 7.86   MAD 0.93

   d. The shoe store is planning to order 500 pairs of Wonder Shoes. Based on the sales in these two stores, suggest how many of each size they should order. Keep in mind that the total has to be 500.

Comparing Two Samples

In this lesson, we compare two populations again but this time using a random sample from each population. The main difference is that now we cannot make inferences about the populations with 100% certainty since the inferences we make are based on the samples and not on the entire populations.

Remember that it is possible, though unlikely, that a random sample would *not* represent the population. A random sample is *likely* to be a representative sample, and therefore any inferences made from it are *likely* to be true—but not 100% guaranteed to be true. Also, we can increase the likelihood of a random sample being a representative sample, and the probability of making true inferences from it by increasing the sample size.

To compare two populations based on two random samples:

1. Check whether the measures of center (mean or median) differ, and by how much.

2. Check whether the data sets differ in their variability. For that, you can use the range, mean absolute deviation (if mean is used for a measure of center), or interquartile range (if median is used for a measure of center).

3. Describe the overlap of the data sets.

Boxplots drawn side-by-side are especially useful for this purpose.

Example 1. Medical researchers are studying whether or not a certain nutrient helps people avoid depression. They choose two random samples: one from among people who don't get much of that nutrient, and the other from among people who get plenty of that nutrient. Then they interview the people using a questionnaire that is scaled from 0 to 10 points. The more points people get, the more symptoms of depression they experience in their lives.

The results of this hypothetical study might look something like this:

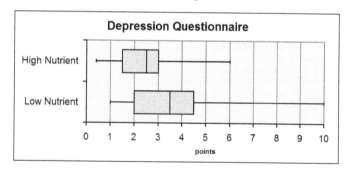

Let's describe the differences and similarities in the two samples:

1. From the medians and ranges it looks like the people who took the nutrient tended to experience fewer symptoms of depression overall than the people who didn't.

2. Both samples show a large variability, but the people who didn't get much of the nutrient vary a lot: all the way from 1 to 10 points, which means some people in the sample don't experience much depression at all, some experience it a lot, and some are in between.

3. Also, we can see there is a large overlap in the distributions. Both groups have people who experience little depression, people who experience some depression, and people who experience a medium amount of depression. However, the people who get a lot of the the nutrient don't experience the highest level of depression (point counts 7-10).

Based on the two samples, it seems this nutrient has some positive effects, but the nutrient alone doesn't totally prevent depression. As is typical for medical studies, the researchers might state in the end, "More studies are needed to examine interactions with other factors."

1. Researchers interviewed families in two cities asking them how much money they spent on food. The boxplots show part of their results.

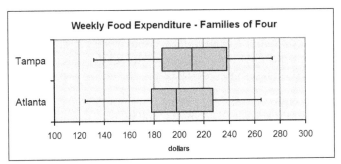

a. In which city (if either) do families appear to spend more on groceries?

b. Estimate the difference in the medians.

c. Comparing this difference to the interquartile range, describe the significance of the difference.

d. Describe the overlap in the distributions.

e. Overall, describe the differences and similarities in how much money families spend on food in these two cities.

f. Let's add an imaginary city called Cheaptown to the comparison. Draw a boxplot for it so that

- the median weekly expenditure on food is low, in the $160s;
- the variability is small: both the range and the interquartile range are only about half of the range and of the interquartile range of the expenditures on food for Tampa and Atlanta.

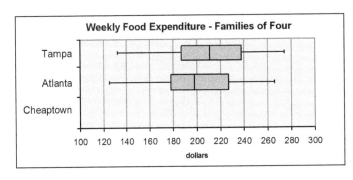

280

2. Researchers studied the texting habits of teenagers in two different schools. They selected a random sample of teenagers from each school. The double (or back-to-back) stem-and-leaf plot shows the number of text messages sent by each teenager in the sample on one particular day.

   a. Just looking at the two distributions, overall which group appears to send more text messages?

   Which group appears to have a greater variability in the number of text messages sent?

   b. Find the range and median for each sample. The interquartile range is given to you.

   School 1:
   Median _____ Range _____
   Interquartile range: 19

   School 2:
   Median _____ Range _____
   Interquartile range: 23

   c. Do these values support your answers in (a)?

Number of Text Messages Sent in One Day

SCHOOL 1 Leaf	Stem	SCHOOL 2 Leaf
	3	5
	4	2
3	5	1
2 1	6	2 6 8
9 9 5 5	7	0 4 4
8 7 3 2	8	2 5
5	9	1
4	10	
	11	5
	12	3
	13	
	14	
	15	

(Note: The stem is the tens digit and the leaf is the ones digit. For example, the first value for School 1 is 53 and the last one is 104.)

3. In this exercise we will study air pollution levels in Chile and Brazil. The data below give the annual mean concentration of fine particulate matter where the particles are smaller than 10 microns (PM10), in micrograms per cubic meter ($\mu g/m^3$) in various cities in Chile and Brazil.

   Brazil 17 21 22 23 25 26 29 30 30 30 31 31 31 31 31 31 32 32 33 34 34 35 35 35 35 35 36 38 38 39 39 42 43 45 46 46 50 52 67 81

   Chile 34 35 36 37 38 38 41 44 45 46 52 52 52 53 58 59 61 62 66 69 69 70 71 79   Source: WHO

   a. Add the data for Chilean cities into the double stem-and-leaf plot.

   b. Which country appears to have cleaner air in its cities?

   c. Determine the medians.

   Brazil: _____ $\mu g/m^3$

   Chile: _____ $\mu g/m^3$

   d. Do the median values support your answer in (b)?

Pollution levels

Brazil Leaf	Stem	Chile Leaf
7	1	
9 6 5 3 2 1	2	
9 9 8 8 6 5 5 5 5 5 4 4 3 2 2 1 1 1 1 1 1 0 0 0	3	
6 6 5 3 2	4	
2 0	5	
7	6	
	7	
1	8	

3 | 4 means 34

4. A bank conducted a study about the waiting times people experience in their branch locations. The boxplots show the results for two particular branches. Based on the boxplots, compare the differences and similarities in these two samples, and make inferences about the total waiting time people experience in these two branches.

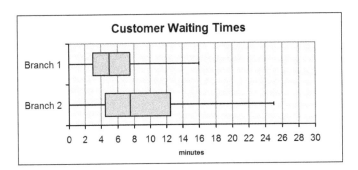

5. A double-bar graph is also sometimes used to show differences and similarities between two data sets. The graph below shows the fuel efficiency in miles per gallon (MPG) for a group of small and standard sports utility vehicles (SUVs).

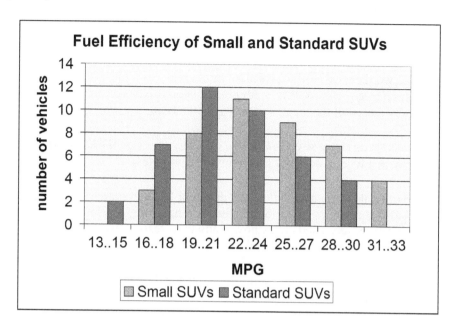

a. Study the graph carefully, and then describe the differences and the similarities in the fuel efficiency of the two classes of cars.

b. Based on what you have studied so far, which type of graph do you feel makes it easiest to compare two data sets: a double-bar graph, side-by-side boxplots, or a double stem-and-leaf plot? Why?

6. Now it is time for you to study and compare two different populations based on random samples from them. Choose from a project listed below or come up with one of your own. Make a double stem-and-leaf plot, side-by-side boxplots, two dot plots, or a double-bar graph to show the results. Then make inferences about the medians and variability in the two populations.

Project ideas

1. Compare the length of words in two different school books. For example, you could compare a 4th grade and a 7th grade science textbook or two 7th grade history textbooks from different publishers. Is there a significant difference in the lengths of words in the books?

2. Compare the traffic on two different streets or at two different times on the same street by counting how many cars pass by in a fixed amount of time (such as 5 minutes).

3. Compare the number of total search results that two different Internet search engines give for a particular topic. For example, conduct searches on recipe-related search terms ("omelet recipe," "easy muffin recipe," "quick vegetable soup recipe," *etc.*) and check how many total search results there are. Use at least 30 different search terms to obtain your sample, and more is better.

   The difficulty here is in obtaining a random sample. To get a truly random sample, you would need to choose randomly among recipe-related search terms that people *actually* use, not make up search terms yourself. However, for the purposes of this exercise, you may come up with the search terms yourself.

4. Compare the physical size of children's story books to the size of young people's novels by calculating the area of the front page. Are children's storybooks bigger (in area) than young people's novels? Think carefully: How will you choose your sample to make it truly random?

Chapter 11 Mixed Review

1. Circle the equation that matches the situation.

   The price of a pair of rubber boots is discounted by 1/5, and now they cost $9.40.

   $p - 1/5 = \$9.40$ $\dfrac{p}{4} = 5 \cdot \$9.40$ $\dfrac{4p}{5} = \$9.40$ $\dfrac{p}{5} = \$9.40$ $\dfrac{5p}{4} = \$9.40$

2. Solve the equation in question (1).

3. A juicer that costs $200 is discounted by 15%. Then it is discounted by 20% off of the already lowered price. Find the discount percentage if the price had been discounted from $200 to the final price in one single decrease. Note: The answer will *not* be 35%.

4. A cylindrical can of sardines has a bottom diameter of 6.6 cm and height of 8.5 cm. Another can has a diameter of 10 cm and height of 5.8 cm.

   a. Calculate the volumes of both cans to the nearest cubic centimeter.

   b. How many percent bigger is the larger can than the smaller one?

5. A house plan has the scale 1 in : 6 ft, and in the plan the house measures 5 ¼ in by 6 ¾ in. What are the true dimensions of the house?

6. The picture shows two pairs of intersecting parallel lines.

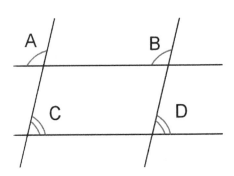

   a. In the picture, mark with a single arc all the angles that are equal to angle A.

   b. Mark with a double arc all the angles that are equal to angle C.

   c. If angle A is 102°, how many degrees is angle C? _____°

   d. What is the name of the quadrilateral that is enclosed by the two pairs of parallel lines?

7. An animal mini-puzzle measures 8″ × 5 3/4″ when finished. How many of those puzzles can fit on a 4 ft by 4 ft table?

8. This is a scale drawing of a toy boat, drawn at a scale of 1:8. Redraw it at a scale of 1:5.

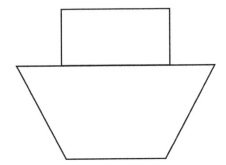

9. **a.** Draw a triangle with sides 2 inches, 3 1/4 inches, and 3 3/4 inches long.

b. Draw a triangle with sides 7 cm, 3 cm, and 3.5 cm long.

10. A spinner with four colors is spun twice.

   a. In the space on the right, make a table, a list, or a tree diagram showing all the possible outcomes of this experiment.

   Then find the probabilities:

   b. P(blue; blue)

   c. P(green; not green)

   d. P(not blue; yellow)

   e. P(yellow or green; red or blue)

Sample space:

11. Tara tosses two coins.

   a. Give an event in this experiment that has the probability of zero.

   b. Give an event in this experiment that has the probability of 1/2.

12. John and Jim decided to check if a particular die was fair or not (in other words, if perhaps it was weighted on one side). They rolled that die 1,000 times. Their results are at the right.

   Based on the results, John said, "Yes, this die is indeed weighted, because we rolled '1' many more times than we rolled '6'."

   Is his conclusion correct? Why or why not?

Outcome	Frequency
1	178
2	160
3	167
4	175
5	167
6	153

13. Sam is studying how well the people in his city like the paintings of the Romantic era. He is planning to stand on a certain street corner near his home and ask passersby if they would like to take part in his study. Explain why his sampling method is biased.

Chapter 11 Review

1. Jake practiced shooting baskets on ten different days. In each practice he shot 50 times, and he recorded the number of baskets that he made. Sadly, he felt he didn't improve any.

Mon	Tue	Wed	Thu	Fri	Sun	Mon	Tue	Thu	Fri
26	33	30	34	29	27	33	27	31	35

   a. Draw a boxplot of the data.

   b. Based on this data, estimate how many baskets Jake would make in 120 shots.

2. Harry belongs to a club that helps abandoned animals. Harry would like for his club to purchase some new equipment, and he wants to find out whether the other members of the club support his idea. Harry plans to interview those club members who stay a little longer after their regular meeting to chat.

   a. Explain why Harry's sampling method is biased.

   b. Suggest an unbiased random sampling method for him.

3. The algebra teacher, Mrs. Riley, had hundreds of test papers to grade and she knew she couldn't finish grading them in one evening. She decided to take a sample of 20 papers from 7th grade students and another sample of 20 papers from 8th grade students to get an idea of how well the students did.

These are the test results for the two samples:

GRADE 7: 56 59 61 64 66 68 68 73 75 75 76 77 79 79 83 84 88 90 96 97

GRADE 8: 52 54 55 58 60 62 62 63 65 68 69 70 72 72 74 77 82 85 92 99

a. Make a back-to-back stem-and-leaf plot of the results.

b. Just looking at the two distributions, does either grade appear to have done better on the test? If so, which one?

Does either grade appear to have a greater variability in its test results? If so, which one?

Test results

SAMPLE 1 GRADE 7 Leaf	Stem	SAMPLE 2 GRADE 8 Leaf

7 | 8 means 78

c. Find the range, median, and the interquartile range for each sample.

Grade 7:   Median _____   Range _____   Interquartile range: _____

Grade 8:   Median _____   Range _____   Interquartile range: _____

d. Do these values support your answers in (b)?

4. The table below shows the results of two separate surveys where students were asked who they would vote for in an upcoming school election.

	Hanley	Johnson	Garcia	Wilson	Evans	Totals
Survey 1	5	25	22	13	10	75
Survey 2	3	23	24	16	9	75

a. What can you infer based on these results?

b. Estimate how many of the school's 1,230 students will vote for Wilson. How far off do you expect that your estimate might be?

5. The boxplots below have to do with prices of eyeglasses in two different stores.

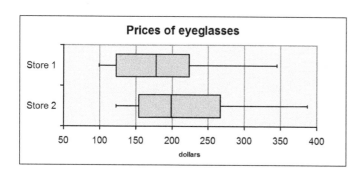

a. Based on the boxplots, does either store appear to have cheaper glasses? If so, which one?

Does either store appear to have greater variability in its prices? If so, which one?

b. Is the difference in the prices significant?

Justify your reasoning.